For all those that appreciate the International Harvester name and what it stands for.

# International Harvester Australia: Geelong Works

*Sarah Tomac*

The information in this book is true and complete to the best of my knowledge. The author or publisher disclaim any liability incurred in connection with the use of this data or specific details. Photographs appear courtesy of Museum Victoria of Spotswood, Victoria and the Author's own collection.
This publication has not been prepared, approved or licensed by CNH Australia. Use of IH logo on cover has been approved by CNH America.



IBSN: 9780646560373

Editor, Designer and Author: Sarah Tomac

Printed by Ingram Books, locations in Australia, UK and USA
Printed in USA

# CONTENTS

**Acknowledgements..............................2**

**Introduction..........................................3**

**SECTION I : International Harvester History**

Chapter

1: International Harvester Company **5**

2: Harvesters' Beginning **5**

3: Competition in Australia **6**

4: International Harvester Company of Australia **7**

5: A New Home **8**

6: Geelong Works—The Ideal Site **11**

7: War Years **13**

8: Expansion **14**

9: Dandenong Works **16**

10: Port Melbourne Works **18**

11: Anglesea Proving Grounds **19**

12: Export Product **21**

13: Beginning of the End **22**

14: The End **23**

**SECTION II: Tractors**

Chapter

1: AW-6 Series **26**

2: AM Series **40**

3: AOS-6 **53**

4: AW-7 **57**

5: AM-7 **66**

6: A-554 **71**

7: A-514 **79**

8: A-414 **84**

9: 564 **95**

10: 434 **104**

11: 686/756 **114**

12: 696/766 **121**

13: 444 **127**

14: 866 **133**

15: 976 **141**

16: 86 Series **149**

17: 86 B Series **159**

18: Acremaster **167**

**Appendix A:** Headers **173**

**Appendix B:** Implements **175**

**Appendix C:** Tractor Build Table **179**

**Appendix D:** Monthly Build **181**

## Acknowledgements


I would like to thank the following people for their assistance in gathering information about International Harvester and Geelong Works. Firstly, I would like to thank David Atley and Kenneth Updike. Without them this book would not have been as complete as it is. David, a former IH mechanic, has provided countless hours of knowledge and information to define Harvester's time in Australia. Kenneth has given me guidance and has been my personal cheer squad—always sending me something just when I needed it most—be it feedback on a chapter or answers to the never-ending questions that I had in the beginning.

I would like to especially thank Lorenzo and Ken at Museum Victoria. They let me search through everything and were of great assistance with anything IH related. In the hunt for unseen photographs and important and obscure dates, they realized that this is a book that needs to be shared with you, the reader.

There are a few other special people that deserve a thank you as well—Robbie, Tom and Jake, all part of the Aerostart Angels—they provide hope to the lonesome tractor left in the fencerow. Also, to R. Devlin for sharing his time and experience of working for IH in the engineering department with me. I would also like to thank my parents, Frank and Pennie Tomac, for the ability to ask them a million and one questions but not needing any answers. All I wanted some days was a set of ears to listen and help make sure we got this 'just right.'

Finally, but not least, I would like to thank you, the reader, for keeping International Harvester alive—through any means possible. International Harvester Australia provided a way of life for many people, from working in the factory to those working in the hot Australian summer sun on tractors. To those that were growing food, feeding livestock, building dams and transporting goods all around Australia and throughout the world using IH products, thank you.

*Sarah J Tomac*

## Introduction

The name International Harvester means a lot to me. It brings back many memories. From an early age I have been involved with tractors and farm life. As a baby, I was buckled up in my 'tractor' seat, riding along with mom working in the fields. In reality it was just a car seat strapped safely to the tractor, but it was my special seat. Family vacations were worked around the farm schedule and the destination always included a tractor show.

Through the Young Farmers program, I went on exchange to Australia, staying on different farms and stations, learning more about agriculture, always finding out which tractor was used the most. As a result, I was able to tour some very memorable collections.

My passion for Harvester and what it stood for led me to write this comprehensive guide about tractors that were built at Geelong Works. While researching the tractors, the story of International Harvester and their presence in Australia was also revealed in the archives. With full specifications about every tractor, I may have gone overboard, but feel that you, the reader, will appreciate this detail.

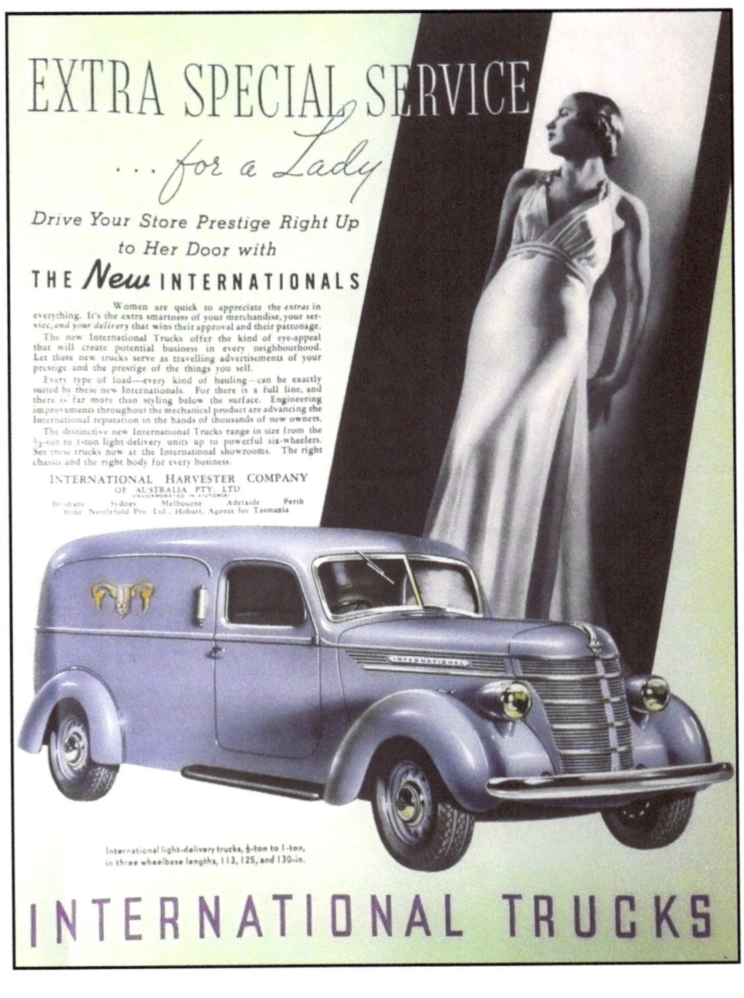


*This advertisement is giving prestige to the products that were carried. Very few advertisements for trucks or tractors were aimed at women—this is as close as one got.*

McCORMICK
MADE BY INTERNATIONAL HARVESTER COMPANY OF AUSTRALIA PTY LTD GEELONG AUSTRALIA
A
OS-6
ORCHARD

## INTERNATIONAL Harvester Company

The invention of the reaper in 1831 opened up a flood of progress, changing the way of farming forever. Between 1830 and 1850, nearly every aspect of farming underwent major changes. The steel plow was introduced which replaced the wood and cast iron implements. In 1834, the first threshing machine appeared, ending the need for laborious hand-thrashing. Hand sowing of seed became obsolete with the introduction of the grain drill allowing rapid, accurate planting. Harrows and cultivators also underwent major improvements, and eagerly accepted by the pioneers of America.

Between 1830 and 1840, the time required to harvest an acre of wheat was reduced from about 37 hours to roughly 11 ½ hours. Physical labor was an even greater decrease.

All of these changes led Cyrus McCormick to build his reaper factory in Chicago in 1846. His business grew in this new location as it was closer to the 'new frontier' of America's farmland in the west.

Cyrus was always thinking about the interests of the farmer and he spent most of his time designing and testing improvements on the reaper. He also insisted on the highest possible standards of quality and workmanship in his machines, always looking for weaknesses and improving on them.

The first major improvement to the reaper was in 1852, adding a self-rake device used to sweep the grain from the platform, reducing another person needed for harvest.
World-wide recognition first began in 1851, when the reaper was entered in the exhibition of industries of all nations at the Crystal Palace in London. The reaper performed in a series of field tests, under all conditions and so efficiently that it received the highest award of the show, the Council Medal. This allowed McCormick to make arrangements with a British firm to build the reaper. Throughout the next decade, the use of the reaper spread rapidly throughout all of Europe. The world-wide market was born.

**Harvester's Beginning**

The first McCormick reaper arrived in Australia in 1852; 21 years after Cyrus Hall McCormick demonstrated his reaper to a skeptical gathering at his father's farm near Steele's Tavern, Virginia, USA.

Two separate organizations, one selling McCormick and one Deering, formed in Australia in 1884. The American merger of these two companies along with a handful of other major manufacturers of agricultural equipment occurred in 1902. In 1903 International Harvester Company of America (IHCA) registered for trading in Australia. In 1904 a joint venture with McCormick and Deering saw the company selling not only their own farm implements, engines and vehicles, they were also sole agents for Buffalo Pitts steam engines, Chattanooga reversible disc plows, Cockshutt plows, Deere plows, Oliver plows, Sanders disc plows and Zealandia milking machines.

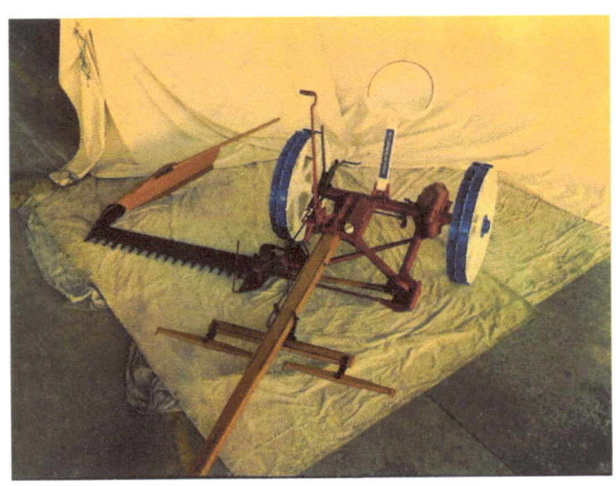

*Middle and Above: GL-9 Horse drawn Mower*

*Bourke Street Premises, circa 1931*

The company headquarters at 545-557 Bourke Street, Melbourne were established in 1904, with 28,000 square feet for agricultural trade. The building sits on the corner of Bourke Street and Church Street, a narrow side street. The first or ground floor is devoted to showrooms, the second floor to offices, and the third floor to storage. The rear of the building is equipped with one 5-ton and two 3-ton travelling cranes, also with an electric hoist for transferring goods to and from the third floor of the front premises.

All International Harvester goods were sold from this building as well as their agency lines, Osbourne Division, the steel ploughs of Deere & Company, as well as the farm engines and thrashers of the Buffalo Pitts Company.

Many other buildings in and around Melbourne became occupied as the needs of the company expanded. Motor-truck assembly, service station, spare parts, showroom and many other aspects to operating a large company were scattered throughout Melbourne and its suburbs.

IHCA assembled most of their components, mainly tractors, in Spotswood, a small suburb just outside of Melbourne, in Hall Street. The company sent these imported implements and tractors all over Australia from this location until Geelong Works was opened in 1939.

**Competition in Australia**

Agriculture equipment has been manufactured in Australia since the very beginning of the agricultural revolution. In 1843 a stripper was developed by John Ridley of Hindmarsh (near Adelaide), South Australia, and in 1885 a stripper harvester was produced by H.V. McKay company of Sunshine, Victoria. Stump Jump disc cultivators and ploughs, combines (seed and fertilizer), as well as header harvesters were all in production by IH competitors by 1917.

*Middle: Tractors leaving Spotswood Station, bound for the Wimmera Region, May, 1927*
*Below, Left: Assembling McCormick-Deering tractors, Spotswood, VIC, September 1923*
*Below, Right: Erecting Harvesting Machines, Spotswood, VIC, September 1923*

*A record consignment of McCormick Deering Tractors at International Harvester's Works in Spotswood, ready for shipment to farmers in the Wimmera Region, Victoria. May, 1927*

IHCA operated quite well on the basis of imported machinery from USA, Canada and later, from Europe. This did not sit well with H.V. McKay, who could not supply his machines at a price competitive to IHCA. In 1904 he accused IHCA of 'doctoring' the import cost of goods so they could pay less import tariffs. The lawsuit lasted from August until December, when the courts found nothing wrong with what IHCA was doing, and ordered the McKay Company to pay all costs.

Some Australian farmers also thought that IHCA was saturating the market with American goods – even though many of the implements were made in Australia by local manufacturers, under the International Harvester name.

Tension was growing in America as well over IHCA success. Farmers and journalists alike, as well as trustbusters, began protesting the monopoly which led to an investigation into the company in 1906. In 1912, the Justice Department filed a suit against IHCA, just one year after forcing the breakup of both Standard Oil and American Tobacco. In 1918, at the end of World War I, IHCA reached an agreement with the US government on the antitrust suit. *

** A Corporate Tragedy, Barbara Marsh*

**International Harvester Company of Australia**

Back in Australia, on January 31, 1912, IHCA sent a memo to 'All Agents and Employees' stating, "In light of the recent press that the Attorney-General of the United States had ordered a dissolution of IHC, we wish to call to your attention that this whole matter is a legal and technical one and you should not imply any wrongdoing on the part of International Harvester or any of its subsidiaries." The memo further states that "the business of this Company in Australia will continue to operate indefinitely, and with greater vigor than before."

It appears that IHCA needed a better sense of 'belonging' in Australia. This was to happen on June 26, 1912, when International Harvester Company of Australia Pty. Ltd. was incorporated with a share capital of £500,000 (about $2,430,000) Australian.

A memo dated June 29, 1912, announced to 'All Concerned' that as of "July 1, 1912, all the assets owned in Australia prior to that date, have been purchased by International Harvester Company of Australia. This includes all the Company real estate, personal property, promissory notes and accounts and its rights and benefits under all contracts, agreements and leases."

*On July 1, 1912, IHC of Australia was born. This check was paid to IHC of America for the purchase of Australian equipment and facilities.*

*Branch Houses for International Harvester; Top Left—Perth, Western Australia; Bottom Left—Adelaide, South Australia; Top Right—Brisbane; Bottom Right—Sydney Photos taken 1939*

The distribution system taken over on the formation in 1912, comprised of a general office in Melbourne, with branches in Sydney, Melbourne, Brisbane, Adelaide, Perth and Launceston with several company warehouses in the main agricultural centers. Agents operating for a commission were appointed throughout Australia. Company salesmen and servicemen were also in direct contact with the public. In 1916 and 1917, the company's warehouses around Australia were sold to agents.

This distribution system was an advantage for the smaller manufacturers that IHA marketed products for. T. Robinson & Co Pty. Ltd. of Spotswood and Gaston Brothers Pty. Ltd. of Melbourne, are two such builders. IHA marketed the Robinson Stripper, and most of the Gaston product line – Header-harvesters, cultivators, disk plows and earth scoops. Although they turned out high quality machines, both manufacturers had buildings in poor conditions, badly equipped shops and had crude factory methods.

Imported trucks usually consisted of the running gear and motor. A local manufacturer, like the Peters, Bryden & Peters Company, usually made the bodies and truck cabs – mostly by hand.

The success of importing machinery lasted until the world-wide economic depression of the 1930s when the Australian government imposed heavy import duties on farm machinery. As this was IH main trade, sales were greatly affected. This resulted in renewed efforts to find and establish IH own factory.

During this period, from 1904 to 1930s, parts for implements as well as whole implements were being made for IH at several different locations. These arrangements created a potential problem for IH as manufacturers were not afraid to copy IH designs and sell them as their own, with few – if any changes. One such company, Mitchell & Company, a small implement plant located about four miles from the center of Melbourne, made Drills, Disk and Moldboard plows, and rakes for IH. Mitchell & Co also sold implements under their own name, one of which was a rake. Their rake was a direct copy of IH with the exception of wheels, which were of an inferior design.

**A New Home**

In January 1937, A team of four IH executives; Mr. McHenry, Mr. Messenger, Mr C.H. King and Mr. Seidenbecker, from Australia and USA,

*Left: Construction of Geelong Works, August 10th, 1938; three weeks after the first sod was turned.*

*Below: A ship coming into Geelong Works, most likely carrying iron ore. (1940)*

scoured the country looking for a suitable place to build the IH factory. Their main focus was in New South Wales and Victoria. The major criteria for a good site were quality roads, rail and water access. They visited many companies to get a better understanding of the market, and to discuss suitable locations.

The different manufacturers visited were overall quite happy for IH to visit and ask questions. Only one manufacturer, John Shearer, tried to discourage IH from expanding the implement business. Other companies, such as Goodyear Tyre and Ford Manufacturing of Australia were more than encouraging for IH to build their factory.

Ford Manufacturing of Australia, now Ford Motor Works, was once in the same position. They even had purchased land in Sydney, near the Parramatta River, but after further research, abandoned Sydney and chose Geelong instead. There was an ample supply of unskilled labor in Geelong, part of it owing to the large number of women employed at the woolen mills.

The four IH executives also visited in Melbourne at length with the Minister of Agriculture Mr. Hogan; the Minister of Land, Mr. Lind; and the Minister of Trade Treaties, Sir Henry Gullett.

They discussed different sites with water access in the neighborhood of Melbourne. A site near present day Holden Motors, in Fisherman's Bend, Port Melbourne, was mentioned but later discarded due to the lack of railroad facilities that would never eventuate.

Vacant land in Spotswood was also discussed. It was the only land left in Melbourne area with water and rail connection. Further examination of the site revealed that part of it had been quarried and out of 40 acres, only half was useable without filling and leveling. The rail switch would be shared with Shell Oil Company and there would also be difficulties with the slope and grade.

Acquisition of the Geelong site was kept secret until after the purchase on September 29, 1937. The "Herald" newspaper headline on September 2, 1937, announced: 'Big Industry for Geelong: Crown Land Sale Announced'.

The article speculates that the industry will be financed from overseas, and that it will employ more regular hands than nearby Ford motor

works. The gazetted land will go to auction on September 29. In order to prevent speculative buying, conditions of the sale were added that a factory was to be built within three years with necessary equipment costing no less than £200,000 ($400,000) and employment given to 200 employees.

The same article goes on to state that the [then] expensive price of £300 ($600) per acre for 45.76 acres had been set for the area known as Fletcher's Freezer Works, a meat processing company that specialized in sheep, which burned down about 1923. The area has deep bay frontage and a wharf, which was built for the old works. Road and rail transport is also established, which meets all the industry's requirements. The price paid per acre was about

*Above: A Train load of steel arrives at Geelong Works, 1940s. Below: Progress of Geelong Works, August 29, 1938*

$200/acre more than Ford motor works paid in 1925.

People around Geelong woke up to the headline on September 30, 1937: ***Another New Industry for Geelong: International Harvester Secures a Site with the purchase of North Shore property***

"An event of the highest importance to the industrial and general development of Geelong, was the announcement of IH intentions to establish an agricultural machinery factory in North Shore, Geelong. The company is a section of the great International Harvester organization which is established in many parts of the world." This announcement was made after the company had purchased the land at an auction sale at the Lands Office, McKillop St, in the morning for the price of £300 ($600) an acre, or a total of £13,730/12/6. ($27,461.25) There was only one bid.

Mr. G.L. McHenry, Managing Director of IH made this statement. "The new plant at Geelong will be designed and equipped for high efficiency to enable it to meet every demand in the production of its products and the commu-

*Geelong Works, 1970s. Land is bordered by the curved road (Seaside Parade) and water. Product Engineering Centre located across the road from Geelong Works, above directional arrow.*

**Geelong Works— The ideal site**

The site chosen for International Harvester was ideal. It was near a sheltered foreshore with a deep channel. Beside this channel is a large jetty, leading directly into the works. At the eastern end of the pier a road runs due north and south through the middle of the works. The highway and a railway station are nearby.

From the air, it appears to be a spreading block of buildings set back from the shore. The buildings are set to allow for expansion. The 46 acre site is an odd shape, with the land high and well drained. The buildings are orientated due north and south, and east and west. The company built a foundry for the manufacture of grey-iron, malleable and non-ferrous castings. A forge shop for all steel parts required was made. The buildings for originating cast and steel parts are located on the eastern side of the main roadway.

The manufacturing buildings were of saw-tooth design, with skylights facing south, to provide even light free from sun glare. It was on the western side of the main road through the works, and received the product from the foundry and forge shops.

On the western side of the main road through the factory was the warehouse building. The east end of this building had an overhead crane that spanned two railway tracks and a roadway, providing crane service for both road and rail transport.

A feature of the factory was the way the various services were installed. All pipes, cables, conduits and such passed through large tunnels between the buildings to allow for safety, extensions and easy maintenance.

nity as a whole will benefit. Not only will Australian materials and equipment be used to the greatest possible extent but allied industries will be considerably assisted as well. The company's plans provide for progressive expansion over a period of years and the 45 acres affords room for considerable future development."

When asked why Geelong was chosen for the factory site, Mr. Messenger, chairman of directors for IHC, explained that many important reasons had helped to make the final decision.

First off, experience in several countries overseas demonstrated the advantage of locating a factory of this kind outside the larger cities and the efforts to start and develop a new industry were appreciated by the local community. Melbourne was unnecessarily crowded with manufacturing industries which created demands for certain classes of labor, had housing difficulties, caused traffic congestion and other problems. Instead of intensifying the situation, a location outside of the Melbourne area with all the necessary requirements and similar facilities for shipping would make an ideal location. Geelong possessed all these requirements and provided an ideal location for both manufacturing and shipping.

The wharf at Corio North was to be extended nearly three times it present length, and the channel deepened for the overseas and interstate ships to be able to tie up right beside the factory – a valuable feature.

Most of the men would be locally employed, as the Geelong district consisted of an ample supply of the unskilled and semi-skilled labour needed for the production of agricultural implements. A small team of certain knowledge and skill for the manufacture of the products were brought in from Canada and the United States, but wherever possible it was the policy of the company to employ the local Australians.

Mr. Messenger also stressed that the company would use as much Australian materials and machinery as possible to construct the new factory.

Construction was soon underway after the first sod was turned on July 19, 1938. Four months after construction started on GW, the first iron was poured at the foundry in December 1938. The plant officially opened May 22, 1939 and employed 216 people.

*Progress of Works, Above—September 14, 1938, Left—November 7, 1938*

Company Headquarters were also relocated in 1939 to City Road, South Melbourne. This consolidated the many buildings that were scattered around the Melbourne area to one location and with convenient access. The new premises, located at 171-205 City Road, were leased and are within half a mile of the main centre of Melbourne.

The move to City road allowed the company's growing motor-truck business to maintain a modern and up-to-date service station with efficiency in layout, equipment and operation. The new premises had showrooms for trucks, tractors and farm machinery. The building also features a ground floor frontage of 450 feet, the finest display space for this type of business in the southern hemisphere. The new site also houses the thousands of spare parts needed to be supplied to people in Victoria, Tasmania, and the Riverina (NSW area). The building is also home to the general Australian office and the Victorian branch office, which handles all the distribution for that State, Tasmania, and the Riverina part of New South Wales. Branch houses are established in the other capital cities, which oversaw sales, finance and service for their respective territories.

*Above: Left, City Road Headquarters; Right, Advertisement during the war*

*Right: K5 International Trucks, in front of City Road building, receiving final inspection by the military.*

**War Years**

Early in 1940, Geelong received its first military contract. The Commonwealth Government had exercised its right to utilize the services of the plant and its work force. Geelong Works became a large contributor to the war effort, manufacturing guns, tanks, shells, aircraft and other war equipment. Geelong Works still continued to produce certain lines of farm equipment to help feed the thousands of soldiers overseas. Engineers at the works were able to develop equipment for large scale vegetable production during the war. As a result, many new lines of vegetable equipment were available in Australia.

*International Harvester played an important part during World War II. In addition to building many parts essential for war, they were also a base for the fairy battle bomber airplanes. In the above photo you can see the white buildings of the army barracks protecting the airfield with planes ready for takeoff. (1940)*

At the end of the war, Geelong Works turned back into an agricultural manufacturer, determined more than ever to provide the Australian farmer with everything needed. The company realized that Australia was a country with huge potential and poised for a gigantic industrial revolution.

**Expansion**

In 1947, plans were completed and construction begun to double the size of the factory, manufacturing for the first time tractors, trucks and power units. Parts of the extensive expansion included a grey-iron foundry, tractor dynometer house, castings cooling area and another special dyno test house for the experimental engineering department. Overhead cranes were installed in the steel yard due to the expansion of raw materials needed to make the ever-increasing range of IH products.

The Head Office building in Melbourne was purchased in 1948, ending a ten year lease. That same year also saw the start of another expansion to the International Harvester line – Trucks. 56 acres were purchased in Dandenong for the manufacture of trucks – from passenger to Army to over-the-road trucking – Dandenong

*Advertisements from 1940s*

became the second IH plant and home to truck manufacture in Australia.

During all this expansion, work continued for the employees building implements, tractors and other equipment. Tractors still were imported from overseas, although the works now had the ability to cast some of their own parts using the dies from USA for these tractors. In order for a tractor to qualify for Australian-made status, a certain percentage of the contents in the tractor had to be locally designed and made.

On April 15 1948, the first Australian made tractor, the AW-6 rolled off the production line. This tractor was soon found in every part of Australia and New Zealand – and working on every imaginable project from farmland to road works and industrial areas.

The formal re-opening of Geelong Works after completion of the building works was scheduled for March 21st, 1949. This grand re-opening was postponed however because of a strike on March 4th, 1949, that resulting in the complete closure of Geelong Works. The disagreement went before the disputes committee of the Geelong Trades Hall Council who declared the Geelong Works closed that afternoon.

The strike was not between the Works and its employees, but was due to 24 clerical workers refusing to join the Federated Clerks' Union because of a flyer that was sent to the non-union members earlier in the year, that implied they would be out of a job if they did not join. On March 4, 120 members of the Union ceased work at International Harvester, which was closely followed by 596 members of the Implement Makers Union going on strike in sympathy with the clerks. The next morning about 100 members of the Amalgamated Engineers' Union also went on strike in sympathy with the clerks.

The strike was so heartfelt that buses and taxis to the Works were banned and members of the Motor Transport Union were on hand to make sure that employees did not report for work.

The managing director of International Harvester Company of Australia, Mr. R.B. Bradley, stated that "it had never been the policy of the company to oppose or discourage membership of a union in any way. It also had not, nor intended, to discriminate against any member of a union because of their legitimate activities in connection with union membership. The company recognized that the desire for union membership or otherwise must remain the prerogative of the individual employed."

Mr. Bradley also stated that "the company made it possible for the executive of the union to speak with the clerks at the factory in their drive for membership, but when their results

were unsuccessful, the members of the FCU went out on strike." On March 14th, 1949 the 25 workers agreed to join the union, and the strike was called off.

Demand for a truck assembly line was so great that the trucks were temporarily produced at Melbourne's City Road headquarters. The first Australian made truck to be mass produced came off a temporary assembly line on August 3, 1950.

*Three views of the Dandenong Works; 1960s*

**Dandenong Works**

On August 4, 1950, a ceremony to begin construction of Dandenong's first major industry, manufacturing motor trucks, was took place in the middle of an empty paddock. Managing Director of International Harvester Australia, Mr. W.W. Killough, turned the first soil.

Construction of the plant at Dandenong moved quickly and June 27, 1952 the official opening ceremony took place – four months after the first trucks begun production at the plant.

In Geelong on September 11, 1952 another great moment was celebrated with the 10,000th tractor built at Geelong Works. A Super AW-6 tractor was driven off the assembly line by the then Minister for Trade and Customs, Sir John McEwen.

*Left: Display days were popular at Geelong Works. They were usually held in May to coincide with the annual dealer meeting. The front lawn would be filled with all types of equipment. Employees and their families as well as the public were invited to inspect the machinery.*

*Below: Advertisement for Harvester featuring AOS-6, Super AW-6, AR truck line and the Countryman refrigerator*

Improvements were continually made to Geelong Works as technology improved and demands increased. With the peak number of employees in 1952 of 2,578, an industrial relations centre was established in 1954 to look after the employee's needs. Soon after, a medical centre was added and a company doctor appointed. The first Australian diesel powered tractor, a Super AWD-6 was made in 1954.

In 1955, the Dandenong Works first expansion program started, the major part was the construction of the cabin sheet metal assembly department.

The increasing manufacturing facilities at Geelong and Dandenong required a continually growing employee workforce to ensure the successful operation of the company. Such numbers were not available in Australia – which was rapidly expanding in all branches of secondary industries. Large scale Government sponsored immigration programs provided a partial solution and Geelong assisted by providing housing, using the former army barracks on site, subsidizing the cost of selection overseas for the tradesmen immigrants and employing many of these successful applicants from Europe in the Works.

Ten acres of land across the road from Geelong Works was purchased in 1955 for future expansion.

The next year, 1956, a comprehensive hot-blast system was installed in Geelong Works grey iron foundry. This included upgrades to the melting section, core room, core sand preparation and

*Left: Announcement of International Harvester's new line for home use; the "Defender" refrigerator (5 ½ cubic foot), June 1946*

*Right: A newer model, at 7 cu. ft., this was aimed for the rural housewife, July 1953*

storage area, pouring and moulding area, mill room and castings handling and cooling area.

Another major upgrade in 1956 was the opening of the new cafeteria which seated 560 people and allowed employees to be served a hot three course meal through one of the four U-shaped serving areas – each capable of serving 100 meals in eight minutes. Mr. H. McKim recalls even if you brought your own lunch, you were expected to eat in the one area, the cafeteria. IHA kept the family atmosphere alive this way.

**Port Melbourne Works**

International Harvester expanded again in 1958 with the purchase of Fowler Engineering Pty. Ltd., a division of Fowler Constructions Pty. Ltd. Possession took place July 14, for the manufacture of construction and earth moving equipment in Australia.

The smallest of the three Works, at just over seven acres, Port Melbourne Works, located at 780 Lorimer St, Port Melbourne, Victoria, is just a hundred yards from the wharf on the Yarra River.

Focusing mainly on small production runs, the Port Melbourne Works assembles or manufactures over 100 types of products. Bulldozers, bullgraders, skid shovels and clamshells, Drott units and other equipment previously made at Geelong Works have been transferred to Port Melbourne. This expansion allows for several new machines such as Hough Payloader and International Payhauler, to be made at the Works. Manufacturing also includes cranes, diggers and loaders previously made by Fowlers. A wider range of Australian made IH products and more employment opportunities has strengthened Harvesters' policy of providing more equipment for Australia's transport, agriculture and industry.

The Port Melbourne Works site was upgraded with buildings, manufacturing facilities, raw material and other general improvements as well an extension and upgrade to the cafeteria. Many of the employees that were employed at Fowler Engineering were welcomed into the Harvester family.

With three manufacturing facilities building products, there were separate engineering departments for drawing and testing of all the products. The need for one location was evident and in

1961, building of the Product Engineering Centre began. This was located across the road from Geelong Works on the ten acres that had been bought in 1955. The facilities included a testing laboratory, drawing office, workshop and prototype assembly area, five dynameters, a stress laboratory, a rig test laboratory and in addition hydraulic and electrical laboratories.

**Anglesea Proving Ground**

Soon it was obvious that a proving ground was needed for testing all the different products. 2,556 acres of land was purchased in 1961 at Wormbete, near Anglesea, Victoria. (About 70 miles south-west of Melbourne)

The proving grounds are one of the most unique in the world. It is one of the only places, that in one area, engineers test and prove for the Australian conditions, motor trucks, tractors, farm equipment and construction equipment. Work quickly started at the Proving Grounds and what better way to test equipment than to build a site with it?

Development of the site was carried out over several years. The first stage included clearing of the heavily wooded land, building a workshop and amenities block a specially enclosed close-security work area. A two and half mile road constructed for the motor truck testing circuit as well as a 2,000 foot long sealed loop road for the tractor testing completes the initial set-up of the Anglesea Proving Grounds.

The first team of IH specialists moved into the test grounds in 1962. Test programs in connection with construction equipment formed part of the work of developing the proving ground. This included sand, gravel, and rock handling, land clearing, dam building, trenching, and road construction. The Engineering Research and Development Centre, across the road from Geelong Works, works closely with the IH Proving Ground.

*From Top: Preparing the Anglesea site firebreaks; Clearing land for workshop site; Hauling bitumen to seal the high speed test track; Inside the workshop.*

In February 1966, two new developments were announced by International Harvester. The first was the introduction of International Harvester Credit Corporation. Incorporated in Victoria on February 15, 1966, with an authorized capital of $6 million, the purpose was to finance the wholesale and retail sales of IH products in Australia. This credit corporation follows the success of the financing established in the United States in 1949 and in Canada in 1959. The IHCC utilized the services of the credit and collection department of IHC which had provided financing services for its dealers and retail customers since the company's incorporation in 1912.

The second announcement was the purchase of 104 acres of land in Werribee, Victoria, for a new farm equipment manufacturing facility. Geelong Works was running out of space to produce products. The greatest diversity of products built by any IH company in the 44 manufacturing locations around the world, comes from Geelong.

The site, located between the old Princes Highway and the new highway, is bounded on the south by the Melbourne-Geelong railway.

The proposed Farm Equipment Works was to have facilities for 551,000 square feet for manufacturing that included machining, sheet metal, welding, fabrication, painting and assembly as well as warehousing area and loading docks. This was to allow for an increase of tractor production at Geelong Works. The Werribee Works was never built. Tractor sales were on the decline just two years later.

An important asset to Queensland customers of International Harvester was the opening of the State Headquarters on March 17, 1966, in Rocklea, Queensland. The design and construction of the 13 and a half acre site on Ipswich Road, 12 miles from Brisbane, took ten years to complete. The imposing view and the quality of the sales, service and assembly operation is evident in the well arranged complex of build-

*Testing at Anglesea Proving Grounds; Above—typical work day March 1978. Below—Scout on a secondary road with bridge crossing, August 1964.*

vice or repair 24 trucks as well as hold sales meetings, the Queensland State Headquarters is able to meet any request promptly and quickly.

A number of projects were completed in 1968 at the proving ground. These included a high-speed truck test loop, eight miles of secondary roads, 34 miles of tracks and specialized testing areas.

The 1968 trend for bigger horsepower tractors resulted in more imports of these machines from IH in USA. The demand for these tractors is evident in the increased acreage farmed by one person, despite a drought.

ings. Two warehouses, a service station, branch sales office, parts distribution centre, and general office area make up the majority of the many buildings. With over one million dollars worth of spare parts in stock and room to ser-

Tractors, Trucks and Farm Equipment were shipped to 90 different countries. The majority of products were shipped to New Zealand and New Guinea.

Products also found a home as far east as Venezuela, Tahiti, Fiji, the Solomon's, New Caledonia, and many smaller islands in the Pacific. They were also shipped to the north and west such as the Philippines and around the coast of Asia from Indo-China to Afghanistan. On the African continent, Egypt, Sudan, Kenya, Uganda, Tanganyika, South and South-West Africa and Angola. Smaller shipments were made to a number of European countries including Denmark, Switzerland and Germany.
In addition to the normal export trade, products were shipped to South and South-East Asia under the Colombo Plan. This was a special plan to help developing countries increase their agricultural output.

**Export Products**

A marketing agreement signed in November 1968, with mechanized sugar-growing equipment manager Toft Bros Pty. Ltd. of Bundaberg Queensland, gave IH a bigger sales potential in the sugar cane growing areas. Toft Bros. manufacture a well-accepted line of sugarcane harvesters and loaders throughout the world.

Under IH policies, the Australian company has no restrictions to the marketing of its products in any part of the world, and Australian-made products are being used in more than 90 countries. The line of Toft sugar cane harvesters and loaders are marketed with the help of International Harvester Export Company, whose headquarters are in USA. The highest sales of this equipment have been to South America, Asia and USA.

*Above: Christmas at Anglesea; Below: Truck Testing Aug, 1974*

**Beginning of the end**

Farm equipment sales for 1969 were down 15.4% on 1968 figures. The downtrend of the country's economy was reflected throughout the entire farm industry. Two reasons for the downturn in economy were the dry conditions and the introduction of the wheat quota system. A general shortage of labor also existed in the Australian manufacturing scene and the company began to hire qualified women in the factories and foundries to overcome the problem. Reorganization of some sections of the company and reassessment of personnel duties were made to assist with the labor shortage.

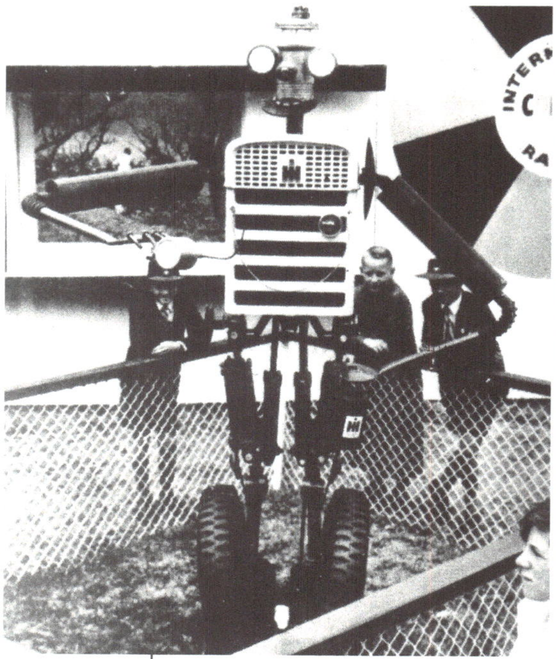

*Children and adults alike admired Robie, the IH robot man, at Sydney's 1961 Royal Agricultural Society Show. Constructed from tractor and truck parts, he flashes his 'eyes,' moves his 'arms,' rotates his 'head' and even carries on a conversation—through a microphone and loudspeaker concealed in his body and connected to a two-way radio.*

On September 1, 1970, in a joint statement made by Toft Bros and International Harvester, Toft and Harvester announced that Toft would resume the distribution of their own line of cane harvesting and handling equipment.

An end of an era was marked on November 1, 1970, when the name McCormick was dropped. The trademark name "McCormick International" was no longer. All IH products from that date forward were marketed with the single name "International," with the exception of the IH Hough and IH Solar Divisions.

Farm equipment sales for the end of the 1970 financial year were down 33% from 1969 and the outlook was bleak. Profits had dropped to the lowest level since 1961. This forced a number of changes at Geelong Works and a reduction was necessary in the number of people employed. With a seasonal turnover of employees, the company decision was not to replace people when they left.

Truck sales were up and as a result, all engine assembly functions were relocated to the Geelong Works to enable greater utilization of space at Dandenong Works.

The introduction of a locally made V-8 petrol engine, made especially for trucks, strengthened Harvester's place in the truck market. Made in Geelong, the V-8 joins the earlier six-cylinder heavy duty engines made there since the early fifties. The assembly of the engines was transferred to Geelong to give greater work support and maintenance of employment, as it was feeling the effects of the depressed rural economy.

*Drilling area at Geelong Works, 1970s*

As a result in the slump in rural activity, Geelong Works was not producing anywhere near the capacity for which it was designed. The rearrangement of manufacturing schedules allowed for more motor truck manufacturing than ever before.

Even with the downturn in rural sales, Geelong Works, in 1974, spent over one million dollars to upgrade to an electric furnace melt centre, to eradicate air pollution. This was part of the company's campaign against air, sea and land pollution. It was to provide the highest clean air standards possible, and improve the overall manufacturing facilities.
The equipment consisted of two coreless induction furnaces of 7.5 ton capacity, and one 32 ton channel type induction furnace. This replaces five cupolas and controls emission of sulfur dioxide, carbon monoxide and nitrogen gasses as well as silica and iron oxides. All this equipment was Australian made.

IHA was converted to a public company in 1974. It was the leading manufacturer and largest single source of capital goods required for essential work in the transportation, agricultural and construction industries in Australia.

Agricultural sales for the next few years were steady, with a small decline in the percentage of overall sales. IHA stayed the market leader in heavy trucks.

*The installation of a 30 ft high sign at City Rd, Melbourne, in 1971, informs motorists of 24-hour service on International trucks. The same service was also offered in Sydney and Brisbane. It was believed to be the only centrally placed revolving symbol in Australia at the time.*

**The End**

IHA had always been given a broad operation plan and the collapse of the parent company had little, if anything, to do with the demise of IHA. There were guidelines, or quotas, to make, which IHA always met. The parent company had reorganization of management three times in four years and that partially reflected in IHA management.

The whole receivership and eventual closure of the company is a slightly confusing story. There were 35 banks involved, management of two companies; International Harvester Australia and International Harvester Credit Corporation (IHCC) a world-wide recession and locally, an Australia-wide drought.

*A two-page advertisement in the 1976 Power Farming Australia Technical Annual catalogue*

The main reason why IHA went into receivership is basically simple economics. IHA continued to produce a lot of trucks, tractors and farm machinery despite the loss of demand in 1981 and the years leading up to it. IHA would sell its product to IHCC, the finance company. IHCC would then sell through dealers to farmers. When demand slowed considerably, IHA was not only left with unsold equipment, it had also sold and profited on tractors and equipment which IHCC had stockpiled, unsold.

Banks had over-lent money to IHA and many of these banks were affiliated with the world-wide reserve of foreign banks. This led to the Australian banks not acting on IHA debts in order to protect the larger loans to the parent company overseas. Initially the banks worked together to keep IHA out of liquidation, but different opinions surfaced and each bank began to protect its own position – with the result of IHA being put into receivership.

IHA and IHCC increasingly drew on credit facilities until the limit was reached. The money market was with-holding more funds to IHA as both companies were fully spent ($300 million)

The company met with the lenders in November 1981, and the plan was simple. They were not asking for more money, just three years grace period for debt repayment. IHA wanted time to ride out the recession and drought. During that time it would scale down the operation, reduce staff levels, reduce stocks and sell surplus assets. This scheme was agreed on between the 35 banks and IHA and was signed on March 2, 1982.

The deal lasted only six months – instead of two years. The failure was in the inability to come to an agreement amongst the banks involved as to how to pay each bank loan back. 47 weeks after IHA asked its banks for help, it was put into receivership – not because of its own problems, but because the banks could not reach an agreement amongst themselves.

The problem with IH was that once IHCC went into receivership on October 2, 1982, the manufacturing company had to follow because both schemes of arrangements between IHA and IHCC were interlocked. The implications were enormous.

One of only two farm equipment manufacturers in Australia, with market leadership in heavy trucks, tractors and headers, IHA was basically selling its goods at cost to compete with the imports. IHA, knowing that it needed to cut its production rate and overheads in order to sell the stockpile of machines in order to survive, took the initiative in December 1982, by closing its Geelong Works. The last cast at Geelong Works was on December 17th, 1982. 800 employees were out of a job by the end of that month.

With Geelong Works closed, the idea was to import machinery and headers as needed as well as sell the extra stockpiled products. Dandenong Works was still fully functional during all this time producing heavy-duty trucks.

In the end, the banks forgave the interest and the suppliers that were owed about $18 million

*Parramatta Rd, Sydney, 1966*

and ignored for 11 months, agreed to cancel half the debt, once half had been paid.

In a complete irony, in 1983, with the drought broken, IHA generated cash faster than predicted in the scheme negotiations and some banks even had funds on deposit from IHA.

An article in *The Age*, June 2, 1983, announces the purchase of the Geelong Works plant, by a state government body, the Geelong Regional Commission, bought the plant after borrowing the money from the State Development Fund. The plant was bought for $5.8 million in a deal that was to re-establish jobs.

A section of the plant was leased to a Geelong based consortium, the Geelong Agricultural Company (GAC). This company was to take over the manufacture of machinery for IHA on July 1, 1983. Part of the agreement over the next five years was that the GAC were to buy all the unfinished products when IHA stopped work, from the Geelong Regional Commission and IHA would take all the products manufactured by GAC.

In March of 1984, IHA was out of the hands of receivership and liquidators. The parent company was still unstable, and IHA was wary – but positive – for the new start.

All the dreams of a new start in the agricultural business ended in June 1986 when Tenneco, the parent company of JI Case, bought the agricultural division of International Harvester in America. This ended 134 years of International Harvester's agricultural presence in Australia. Dandenong Works is still operating today, producing trucks.

# Chapter 1
# AW-6 Series

The first Australian-built tractor to come from Geelong Works rolled off the line on 15 April 1948. Before this tractor, all models were imported from overseas and assembled first at Spotswood, then at Harvester House in Melbourne and finally tractors were built in Geelong, Victoria.

In order for a tractor to be considered Australian built, a certain percentage of the tractor needed to be built or cast in-house. The first Australian built tractor was modeled off the American W-6 which was already being sold in Australia as early as 1942. The American built W-6 was made from 1940 to 1953.

In July 1948, the Australian-made Farmall M expanded the line of locally made tractors. A short four years later, on 11 September, 1952, the 10,000 tractor, an AW-6, came off the production line at Geelong Works.

Identification of the early Australian-made W-6 tractors feature a cast iron oil pan and a deeper fan cowling. Early models also do not have the 'A' prefix. The Super AW-6 tractors have a channel iron frame, which is the same as the AM tractors. This was done partially to keep costs to a minimum, and partially to help with the fit of different implements to both the W-6 series and M series of tractors.

The AW-6 is the perfect all-round tractor for Australian farmers in 1948 with a 4 cylinder AC-248 engine with 38 max belt horsepower, 1450 rpm, spring loaded clutch, external contracting foot brakes, five speed transmission, Belt pulley speed of 899 rpm, and electric start.

*W-6 tractor, first one built at Geelong Works, April 15, 1948. This tractor is missing headlights and mounting brackets.*

**Specifications**

**Engine**: AC-248

4-cycle, 4 cylinder; magneto ignition; kerosene fuel

| | |
|---|---|
| Rated engine speed: | 1,450 rpm at full load |
| Bore and stroke: | 3 ⅞ x 5 ¼ in. |
| **Horsepower:** | |
| Max Belt: | 38.5 |
| Rated Belt: | 32.7 |
| Max Drawbar: | 34 |
| Rated Drawbar: | 25.5 |

Belt Pulley: 11 in diam. X 7 ½ in wide; 898 RPM

Power Take Off: 1 ⅜ in. spline shaft; 537 RPM.

**Forward Speeds:** 1.70, 3.41, 4.45, 5.40, 16.02 mph

| | |
|---|---|
| Reverse speed | 2.07 mph |
| Front wheel tread | 46 ¾ in |
| Rear wheel tread | 53 in |
| Wheelbase | 76 in |
| Total length | 124 ⅜ in |
| Total width | 66 ⅜ in |
| Height, to top of air cleaner cap | 70 ⅛ in |
| Turning radius | 11 ft 6in |
| Clearance, under front axle | 15 in |
| Drawbar vertical adjustments | 12, 15, 18in |
| Drawbar lateral adjustments | 18 ¼ in. |
| Weight: | 5,500 lbs |

on 13.50 x 28 tires including wheel weights

**Approximate Capacities**

| | |
|---|---|
| Fuel Tank—Kerosene | 17 gallons |
| Petrol Tank—for starting only | 5.6 pint |
| Water Cooling System | 5 gallons |
| Crankcase Pan | 6.4 quarts |
| Transmission and Differential case | 10.4 gallons |
| Air Cleaner Oil Cup | 1.8 pints |
| Steering gear housing | 1.2 pints |

Gear and speeds are same as American W-6 tractor with low low gear on 13.5 x 28 tyres

Pneumatic tyre sizes: Regular Rear, 12.75 x 28; Front, 6.00 x 16. Special Rear, 13.5 x 28

*Above: AW-6 tractors and cultivator drills at Charlton, Victoria.*
*Below: Advertisement for AW-6 Tractor*

*Final Inspection of Australian W-6 tractors*

**Tractor Build Data**
1948-1953
Price in 1952 £1,304 ($2,609)

7,409 units built

**Financial Year**

| **July 1—June 30** | **Units Built** |
|---|---|
| 1947/1948 | 266 |
| 1948/1949 | 738 |
| 1949/1950 | 1316 |
| 1950/1951 | 1951 |
| 1951/1952 | 2250 |
| 1952/1953 | 888 |

Exact serial numbers for the AW-6 are unknown at this stage. There is a possibility that the American made W-6 tractor serial numbers precede the Australian-made W-6 tractors. What is known is that the first W-6 tractors off the line have the prefix 'WBK' -same as the American tractors. The Australian prefix 'AW6' does not seem to appear until the Super AW-6 tractor.

## Is it American or Australian built?

## Identifying a W-6 tractor

The early Australian built W-6 tractors and the American built, Australian assembled, W-6 tractors are nearly identical. There are many differences—here are a few quick and easy identifying marks.

**Australian Built Tractor**

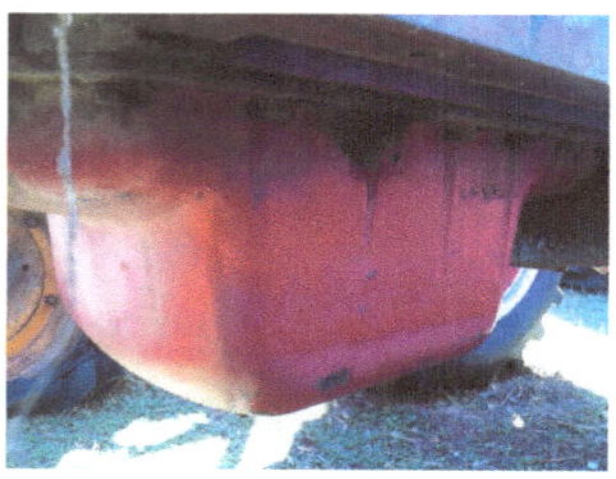

From Above Left: Front emblem is a simple 'IH'; Fan cowling is deeper; Above, Bottom Left: Cast iron oil pan; Painted aluminum tag reads "W6 WBK538 W3" [Engine RPM] 1595

**American Built Tractor**

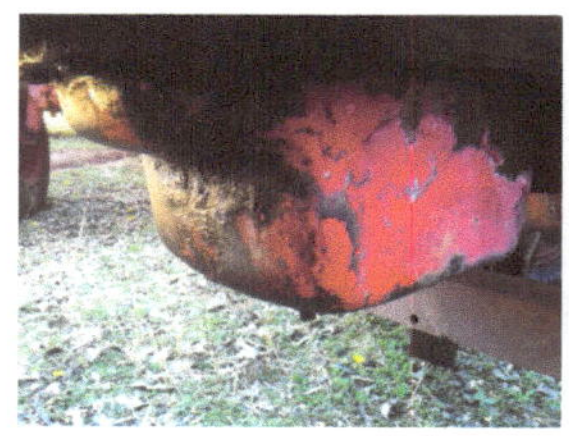

From Above Left: Front emblem is "McCormick-Deering"; Fan cowling is narrow; Above, Bottom Left: Pressed tin oil pan; Etched tin tag reads "W6 WBK11936" [Engine RPM] 1595

| | Australian | American |
|---|---|---|
| **Grill Emblem** | 'IH' emblem | 'McCormick' emblem |
| **Fan Cowling** | Deep | Narrow |
| **Oil Pan (Sump)** | Cast Iron—from AW-6 onwards. Early W-6 have pressed tin. | Pressed Tin |
| **Serial Number Tag** | Aluminum (Painted—see photo example) | Tin (Etched—see photo example) |

*Early Australian built W-6 tractors have a pressed tin oil pan, the Australian cast iron oil pan appears with the introduction of the AW-6 tractor. All parts of these two tractors are interchangeable, and often farmers 'made-do' with the parts that were available, resulting in a tractor that may have conflicting identifying marks.*

The tractor used for identifying the Australian differences is an early Australian built W-6 model. It is serial number WBK538, found by David Atley for the Tom Horwood Collection and Museum.

*The 10,000th Australian made tractor, an AW-6, being driven from the assembly line by the Right Honorable John McEwen, who was then Minister for Trade and Customs. 11 September, 1952*

*Adverts for Super AW-6 tractor with matched equipment*

**Super AW-6**

On 4 February 1953, the first Super AW-6 was undergoing testing trials pulling a GL200 header in a barley crop. The Super AW-6 was released later in the year, 29 July, 1953 in Atherton, QLD.

Not only is the Super AW-6 a new tractor in its power range, but it has many features new to McCormick International tractors.

- New AC-264 engine
- New aluminum alloy, low expansion pistons for better heat dissipation and less vibration
- New distributor for better ignition control
- New tractor chassis frame with attaching points for direct mounted equipment
- New heavier ribbed rear frame
- New two-piece rear frame cover ensuring simple installation of three point linkage or inspection of final drive
- New transmission with stronger countershaft, heavier gears and greater capacity bearings
- New tyre sizes giving better traction and better balance
- New demountable steel rims.

As well as 21% more drawbar power the Super AW-6 has more of everything the tractor operator needs.

- More power and economy from the new larger displacement engine designed exclusively for Australian power kerosene
- More cooling capacity in the larger radiator
- More speed in each operating gear
- More durable gear wear from new hardening process
- More traction from new wide base rear tyres
- More strength and serviceability built into new tractor chassis
- More convenience of operation with full hydraulic systems for 3-point linkage of direct connected equipment and hydraulic remote control of trailing machines.
- More comfort for the driver with relocated seat and foot pedal controls
- More versatility with direct connected or trailing equipment
- More profit per acre with a full line of matched equipment; matched in size, weight and speed of operation to the new Super AW-6 tractor.

**Built to excel** Designed and built in Australia to excel on Australian fuels, which differ from those in other countries, the Super AW-6 has greater power and faster field speeds.

In the ever-present development of mechanization, more acres can be ploughed, more acres tilled and more acres harvested by the Super AW-6 for less cost per unit of product. With remote control or three-point linkage available, work can be done with less effort.

*This illustration show the layout of the controls and gauges of the Super AW-6 tractor.*

The extra comfort and ease of handling of this well-balanced tractor make work in the field with the Super AW-6 a pleasure, not a task. The new AC-264 engine in the Super AW-6 has been exhaustively tested to suit the particular octane rating of Australian fuel. The particular compression ratio and design of the cylinder head have come only as the result of exhaustive laboratory and field testing.

Balanced tractor power is the co-ordination of power, weight, traction and fuel economy. These four factors have been combined by International Harvester's engineers to make the Super AW-6 the best balanced tractor of its power range.

The power of the engine is balanced with the distribution of weight on front and rear wheels and also with the rear wheel traction to produce power to excel on the drawbar. The smooth output of increased power from this balanced tractor means more horsepower hours per gallon and more acres worked for every gallon of fuel.

The Super AW-6 incorporates a host of improvements not seen in its predecessors.

- Improved combustion with complete detonation control
- Improved design of engine especially for Australian power kerosene
- Improved carburetion with better mixture control
- Improved clutch springs
- Improved larger diameter crown wheel and pinion
- Improved crankshaft – larger, to convey greater power
- Improved differential gear surface treatment
- Improved drag link for better safer steering
- Improved hydraulic pump, inbuilt in the changed chassis design.
- Improved clutch serviceability and transmission coupling
- Improved drawbar design to facilitate removal for three-point linkage.

**Standard Equipment**

AC-264 four cylinder 4in x 5 ¼ in engine; variable speed governor controlled from seat; manifold heat control valve; replaceable cylinder sleeves; molybdenum-iron alloy exhaust valve seat inserts; 6-volt battery with voltage regulation; battery ignition; electric starting; oil-bath type air cleaner; fuel strainer; floating oil screen; replaceable element oil filter; spark arrestor and vertical exhaust; adjustable radiator shutter and heat indicator; enclosed external-contracting. Steering foot brakes, operation on the differential shafts; spring loaded foot clutch; front drawbar; vertically adjustable rear drawbar; swinging rear drawbar; adjustable, upholstered waterproof seat; 7.50 x 16 front tyres; 14.00 x 30in rear tyres; 2 front wheel weights; 2 rear wheel weights.

**Special Equipment**

Electric lighting – two front and one rear lamps; 11 in x 7 ½ in belt pulley attachment; rear power take off; hydraulic pump unit; three point linkage – British standard No.2 dimensions; hydraulic remote control unit; rear wheel weights.

**Specifications**

Engine: 4 cyl. Vertical OHV, replaceable sleeves
Displacement: 264 cubic inches
Rated engine speed: 1,450 rpm at full load
Bore and stroke: 4in. x 5 ¼ in.
Horsepower: 45 b.h.p. (Aust. kerosene)
Pistons: Low expansion aluminum alloy
Piston Rings: Three compression, one oil
Electrical equipment:

Electric starting, 6-volt battery, automatic spark advance
Engine lubrication: Full pressure
Cooling system: 4 rows tubes, pressure type
Fuel system: 17 ½ gal kerosene; ¾ gal petrol
Belt pulley; diam. and face 11in. x 7 ½ in.
Belt pulley speed 899 rpm; 2,588 ft/min
Power take off 537 rpm
Steering enclosed worm and gear type
Clutch 11in single plate, Heavy duty springs
Forward Speeds: 1 ⅝, 3 ½, 4 ⅝, 6 ⅜, 15 ⅝ mph
Reverse speed 2 ⅛ mph
Front wheel tread 46 ¾ in
Rear wheel tread 55in
Wheelbase 76 ½ in
Total length 125in
Total width 73 ¾ in
Height, to top of air cleaner 71 ¼ in
Turning radius 11 ft 3in
Clearance, under rear axle 15 ⅛ in
Drawbar ver. adj. 12 ¾, 14 ¾, 16 ¾, 18 ¾, 20 ¾
Drawbar lateral adjustments 18 ¼ in.
Operating weight without operator 7,068lbs

**Tractor Build Data Super AW-6**

| **Financial Year (July 1—June 30)** | |
|---|---|
| **1953-1958** | **2,257 Units built** |
| 1953/1954 | 1137 |
| 1954/1955 | 663 |
| 1955/1956 | 250 |
| 1956/1957 | 148 |
| 1957/1958 | 59 |

Exact serial numbers for the Super AW-6 are unknown at this stage.

Australian Tractor Test number 27, July 1955

*FAO Farm Machinery Advisor, Mr. J. Wylie, training locals inhabitants on a Super AW-6 tractor and GL-132 cultivator drill, operating at 8,500 feet on the Mount Tabina plateau.*

Santa comes to Harvester House on a Super AWD-6 tractor in 1956, delivering toys to the Harvester children.

**Super AWD-6**

- **Australian Diesel** 50 hp, AD264 diesel engine, made in Australia
- **Quick, All-Diesel Starting** Glow Plugs and heavy duty, solenoid-operated starter motor assure easy starting under all conditions.
- **Complete Combustion** Swirl-type pre-combustion chambers assure complete fuel combustion and minimize "diesel knock"
- **Less Fuel per Horsepower Hour** Design of engine and fuel injection equipment provides efficient use of combustible energy
- **Balance** Balanced power with balanced weight, traction and stability, provides smooth, economical operation
- **Trouble- Free Diesel Operation** Low injection pressure with single orifice injection minimizes fuel pump and injector maintenance and provides longer life.
- **Sensitive Governing** Sensitive, full-speed range governor maintains constant engine speeds under varying loads
- **Lugging Power** An exclusive torque control device increases the ability of the engine to "hang-on" without stalling under extreme loads
- **Versatility** Three-point linkage and/or hydraulic Remote-Control the Super AWD-6 is adaptable to all jobs in Australian agriculture

McCormick International Super AWD-6 tractor equipped with 7.50 x 16 front and 14 x 30 rear tyres and electric lighting equipment.

The development and manufacture in Australia, by International Harvester, of a diesel-powered tractor, introduced 22 October, 1953, fills a long awaited requirement. The Super AWD-6 tractor is the first Diesel to be manufactured in Australia for International Harvester.

This versatile tractor, the McCormick International Super AWD-6, introduces a further contribution to Australian Agriculture. Its 50 horsepower engine (the AD-264) provides ample power plus economy for the requirements of most of Australia's farms. Its power is well balanced with the well-selected field speeds and the traction provided by wide-base tyres. Comfort and ease of handling are completely assured by the newly positioned seat and controls. Three-point linkage and two-way hydraulic Remote Control give the tractor driver an all-round-control from the tractor seat, of both direct-connected and trailing implements.

**Strength the Keynote of Construction**

The strong "I-beam" type front axle characterizes the strength that is built into the general construction of the Super AWD-6. The rear frame housing is heavily ribbed to withstand strain, whilst the strong channel-type front frame has attaching points for direct-connected equipment.

**Positive Traction**

Positive traction is afforded by wide based tyres. This reduces wheel spin, a limiting factor to many tractor capacities.

**Positive Power Transmission**

The action of the 12-inch heavy duty clutch is positive and its control smooth and effortless. Heavy duty transmission conveys power effectively to the drawbar.

**Ideal in Rough Going**

With a turning radius of 11ft. 3 in. the Super AWD-6 has a high degree of maneuverability for a tractor of its size and power. It combines a low centre of gravity with the excellent clearance of 15 1/8 inches. This makes it an ideal tractor in rough going.

**Rear Standing Platform**

When not equipped with three-point linkage, the tractor may be fitted with a rear standing platform, which is available as an attachment.

The forward end of the platform is attached to the axle housing and the rear end is supported by braces to the rear fenders. No additional holes are required on the tractor to fit the attachment.

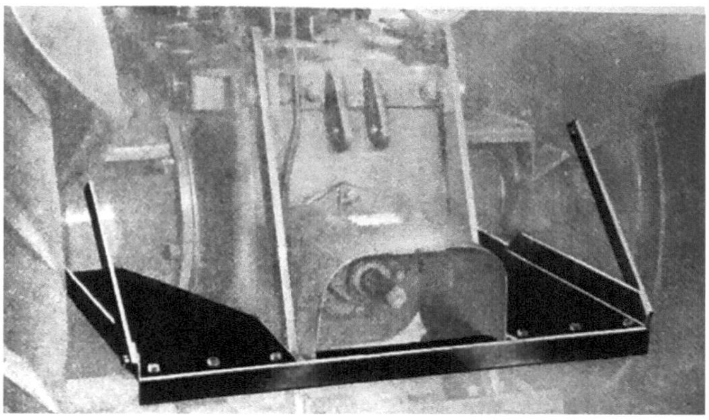

**Regular Equipment**

AD-264 four-cylinder diesel engine with 4-in. bore and 5 ¼ -in stroke. Diesel injection pump with four 8mm plungers, manual primer pump, excess fuel device, centrifugal governor, automatic torque control and single orifice injectors; glow plugs incorporated in nozzle bodies; two 6-volt, 17 plate batteries voltage regulator; electric starting with solenoid-operated heavy duty starter motor; replacement cylinder sleeves; 7-in oil bath air cleaner; replaceable cartridge type fuel filter; water trap fuel filter; floating oil screen; replaceable element oil filter; spark arrester and vertical exhaust; heat indicator; oil pressure indicator; enclosed external-contracting, steering foot brakes, operation on the differential shafts; spring loaded foot clutch; front drawbar; vertically adjustable rear drawbar; swinging rear drawbar; adjustable upholstered waterproof seat; 7.50x 16-in front tyres on demountable rims; 14.00-x 30-in wide base rear tyres on demountable rims; 2 of APWW-507 front wheel weights; 2 of APWW-529 rear wheel weights.

**Special Equipment**

Electric lighting – two front and one rear lamps; 11-in x 7 ½ -in belt pulley attachment; ASAE-SAE Standard rear power take-off; hydraulic pump unit;' three point linkage – British Standard No.2 dimensions; hydraulic Remote Control unit; APWW-530 rear wheel weights. Extra front wheel weights. Rear standing platform.

**Specifications**

**Horsepower**

belt, max.: 47
drawbar, max.: 42

**Engine:**

4-cyl vertical overhead valves
Displacement: 264cu. in.
Rated Engine Speed: 1,450rpm at rated load
Bore and Stroke: 4"x 5 ¼"

Pistons: low expansion cam ground aluminum alloy
Piston Rings: 3 compression, 2 oil

Injection: CAV pump with four 8mm plungers
Injection Nozzles: Single orifice injecting into pre-combustion chamber
Injection Pressure: 700 to 710 lb. per sq.in.

Starting: Direct on Diesel fuel

Starter Motor: Positive engagement, solenoid operated, 4 7/16" diameter

Starting Aid: Glow plugs in nozzle bodies, operated by two position switch

Compression Ratio: 16.5 to 1
Electrical Equipment: Two 6 volt 17-plate batteries connected in series
Engine Lubrication: Full Pressure
Cooling System: 4 rows tubes, pressure

Belt Pulley – diameter and face: 11" x 7 ½"
Belt Pulley – Speed, ft per min: 899 rpm, 2,588'
Power Take Off: 537 rpm

Steering: Enclosed worm and gear drive
Clutch: 12" single plate

Forward speeds mph.: 1 ¾, 3 ⅝, 4 ¾, 6 ⅝, 16 ⅛
Reverse speed, mph: 3 ½

Front wheel tread: 46 ¾"
Rear wheel tread: 55"
Length: 125"
Wheelbase: 76 ½"
Width overall: 73 ¾"
Height to top of air cleaner: 71 5/16"
Turning Radius: 11' 3"
Clearance under rear axle: 15 ⅛"
Drawbar height adjustments: 12 ¾, 14 ¾, 16 ¾, 18 ¾, 20 ¾ -in.
Drawbar Lateral Adjustments: 18 ¼"

*Sixteen Indian farmers, in Australia to study agricultural trends, visited Geelong Works to see farm machinery being manufactured. Here one of the groups gets the 'feel' of an Australian made Super AWD-6 tractor.*

**Tyres**
Front: 7.50 x 16
Rear: 14x 30

**Operating weight**
without operator: 7,298 pounds

**Tractor Build Data**

**Financial Year (July 1– June 30)**
Built from 1953-1958 3,668 Units built

| | |
|---|---|
| 1953/1954 | 745 |
| 1954/1955 | 1,255 |
| 1955/1956 | 924 |
| 1956/1957 | 679 |
| 1957/1958 | 65 |

Exact serial numbers for the AWD-6 are unknown at this stage

Australian Tractor Test number 26, June 1955

# Chapter 2
# AM Series

## AM

The Farmall 'AM' tractor is Australia's first factory produced row crop tractor. Introduced in 20 July 1948, at a Victorian Dealer Conference, It replaces the Imported American built Farmall 'M.' The American Farmall 'M' and Farmall 'H' were offered together during the war period until the Australian plant was able to do more castings. These tractors were a skid unit, with Australia's own designed and cast wheels. The Farmall 'AM' is popular with the cane and tobacco farmers for it's high visibility and row clearance for front or mid-mounted attachments.

There can be some confusion identifying the early Australian made 'M' tractors. The decals on the tractor simply state 'M' and the serial numbers start with the prefix FBK– same as the American made tractors. The Super AM serial prefix may have AM—without a 'Super' on the 'AM' decal. Some early Super AM tractors have 'SAM' decals, but only AM on the serial number tag. It is indeed a Super AM tractor if it has the AC-264 engine. The Diesel tractor had the decals 'AMD' and sports the AD-264 engine. The AM tractor was available only in Kerosene, and used the AC-248 engine.

**Horsepower**

| | |
|---|---|
| Max belt, | 38.5 |
| Rated belt, | 32.8 |
| Max drawbar | 34.5 |
| Rated drawbar | 25.9 |

Engine:
4 cycle, 4 cylinder; 3 7/8 in bore; 5 ¼ in stroke; 950-1450 RPM; magneto ignition; kerosene fuel

Belt pulley: 11 in diameter x 7 ½ in wide; 899 RPM

Power Takeoff: 1 3/8 in spline shaft; 537 RPM. Height above ground level with 11 x 38 tyres 26 ¾ in. end of shaft 14in. ahead of drawbar pin. Shaft central between wheels. Clockwise rotation.

Pneumatic tire sizes: Regular Rear, 11 x 38; Front 6.00 x 16, Special Rear, 13.50 x 32

Regular speeds with 11 x 38 tires: 1.7, 3.43, 4.19, 5.09, 16.11 MPH, Reverse 2.02 MPH

Turning radius 8ft 5 ½ in.
Wheel base 88 ½ in.

Overall dimensions:
Height 64 ½in
Width, 101in
Length 133in

Weight: 5,500 lb on 11 x 38 tyres, with wheel weights

**Tractor Build Data**

**Cost in 1952** **£1,278 or $2,556**

**1948-1953** **3,341 Units built**

| **Financial Year (July 1—June 30)** | **Units** |
|---|---|
| 1948/1949 | 200 |
| 1949/1950 | 700 |
| 1950/1951 | 1,000 |
| 1951/1952 | 1,001 |
| 1952/1953 | 440 |

Exact serial numbers for the AM are unknown at this stage.

# Is it American or Australian built?

## Identifying a M tractor

The early Australian built M tractors and the American built, Australian assembled, M tractors are nearly identical. There are many differences—here are a few quick and easy identifying marks

*Early Australian built M tractors have a pressed tin oil pan, the Australian cast iron oil pan appears with the introduction of the AM tractor. All parts of these two tractors are interchangeable, and often farmers 'made-do' with the parts that were available, resulting in a tractor that may have conflicting identifying marks.*

| | **Australian** | **American** |
|---|---|---|
| **Grill Emblem** | 'IH' emblem | 'McCormick' emblem |
| **Fan Cowling** | Deep | Narrow |
| **Oil Pan (Sump)** | Cast Iron—from AM onwards. Early M tractors have pressed tin. | Pressed Tin |
| **Serial Number Tag** | Aluminum (Painted—see photo example) | Tin (Etched—see photo example) |

### Australian Built Tractor

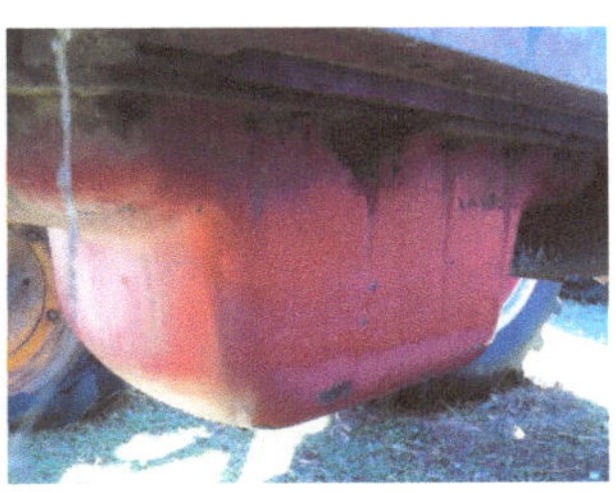

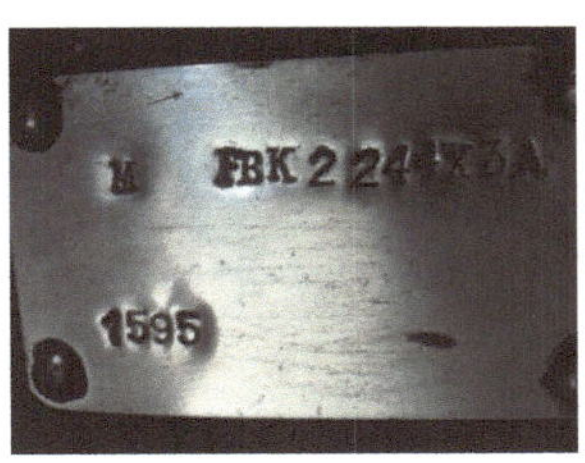


From above Left: Front emblem is a simple 'IH'; Fan cowling is deeper; Above, Bottom Left: Cast iron oil pan; Painted aluminum tag reads "M FBK2244X3A" [Engine RPM] 1595

### American Built Tractor

From Above Left: Front emblem is "McCormick-Deering"; Fan cowling is narrow; Above, Bottom Left: Pressed tin oil pan; Etched tin tag reads "M FBK86956" [Engine RPM] 1595

Right: A well preserved example of the Geelong Built Serial Number Tag. This is an example of an A-554 Kerosene tractor, made from 1961– 1963.
*Tag reads Model "A554K" Serial No. "A554K 3638" Made in Australia by International Harvester Co. Geelong Victoria "1990" RPM Max. Idle Speed. Do Not Overload.*

## SAM

**Farmall Super AM Tractor – The Greatest Value Ever Offered**

The McCormick International Farmall Super AM leads the farm tractor field to new and higher standards of performance. Introduced on 28 July, 1953, the Super AM is the most powerful kerosene Farmall ever built, with 18% more power than the Farmall AM.

The Super AM has more "wanted-and needed" features than any other tractor of its size. Farmers will appreciate the high clearance, the economy of the Super AM, both in operation and maintenance, as well as the ease with which it is handled in all soil conditions. The Super AM leads the field in work output, fuel economy and overall performance.

Adjustable wheel treads enable the tractor to fit any crop, whether it is on broad acres, row crops of maize of sugar cane or mixed farming. The operator sits on a comfortable seat, well up out of the dust zone with uninterrupted vision ahead for straight cultivating and harvesting. All controls are in convenient reach of the driver's seat.

**Reliable Kerosene Engine**

**Every engine tested** Every Farmall Super AM engine is tested for power and then the moving parts are "lapped in" before installation in the tractor. This is the final check among many tests made during every step of manufacture. This ensures that every Super AM engine delivers its quoted horsepower.

**Large Displacement** The Super AM gains power and economy from the large displacement AC-264 engine, designed, field tested and manufac-

**General Features**

- High clearance under the rear axle housings for row crop cultivation
- Strong counter-shaft, heavy gears and high capacity bearings for long service power transmission
- Ideal power-weight balance with new tyre sizes
- "Lift-All" cylinders give selective or delayed lift
- Double-acting Remote-Control cylinders
- Available with full hydraulic system

**Engine Features**

- Engine designed exclusively for Australian kerosene
- Efficient combustion chamber design gives more complete burning of fuel
- Hardened Molybdenum –iron alloy exhaust -valve seat inserts
- Distributor with automatic-advance-control
- Variable speed governor allows speed required to be selected in any gear
- Large capacity pressure cooling system, fitted with thermostat and radiator shutter control.
- Pressure lubrication, with a replaceable element oil filter.

tured in Australia to operate on Australian power kerosene.

**Heavy Duty Clutch Springs** to meet the increase in power of the Super AM, the single plate clutch is equipped with heavy-duty clutch springs. This ensures positive, smooth take up on starting, and non-slip operation at all loads.

**Sturdy transmission and Final Drive** The sturdy transmission and final drive is another example of high quality transmission design and construction. Strong counter shaft, heavy gears and high capacity bearings combine to ensure trouble free long life service. Gears and pinions are drop forged of alloy steel, precision machined, carburized and heat treated to leave them extremely hard on the contact surfaces and relatively soft in the centre to absorb sudden stresses. Gear tooth contours give maximum contact area under full load.

**Clearance, Convenience and Adaptability** The 26 ½" axle housing clearance of the Super AM makes this the ideal tractor for row crop cultivation. This clearance enables the farmer to cultivate a tall row crop to a later stage of growth without fear of damage to the plants.

**Positive Traction** new tyre sizes assure positive traction, to use the extra power in the Super AM engine for heavy drawbar work. The Super AM has an extremely wide range of rear wheel tread adjustment, combined with high under axle clearance for cultivation all kinds of row crops.

**Easy to Maneuver in Row Crops** The Farmall Super AM has only an 8'3" turning circle with steering brakes applied and 8'6" without steering brakes – Less than its own length! This great maneuverability is indispensable to row crop farmers for short turning. Steering wheel position and steering gear mechanism give easy, ready response and result in shorter headlands and maximum time saved on any farm.

**Convenience of Operation** The Super AM is ideally suited for the requirements of open field or row crop cultivation. The front channel frame with slotted holes for quick-attachable clamps

The Super AM tractor fitted with 12 x 38 wide base, 6-ply rear tyres and 6.00 x 16 6-ply front tyres has 36.7 horsepower (maximum) at the drawbar.

provides variable attaching positions for tractor mounted cultivators. Tapered mounting holes are also located on clutch housing and rear axle housing. The quick attachable drawbar has a wide range of lateral and vertical adjustments for draft control of trailing equipment. Power is also available at the belt pulley and the Farmall design provides a clear line of belt drive. Rear power take-off and side drive for planters and fertilizers are also featured.

**Mounted Fertilizer and Planter Attachments** A side power drive operated fertilizer and planter attachments, giving accurate control of the sowing of seed and the application of fertilizer.

**Full Hydraulic System** The Super AM is designed with a full hydraulic system. Pressure supplied by the famous Lift-All pump, gives convenient and positive control of direct mounted equipment, three-point linkage equipment and Remote-Control of trailing equipment.

**Extra Low-Low Gear** The Super AM has a low-low gear for the accurate planting of row crops, a typical example being sugar cane. This low-low gear is also invaluable for the operation of power take off machinery in heavy crops.

**Hydraulic Three Point Linkage: For finger touch control of direct connected equipment**

The Super AM is designed not only for the hard, tough pulling work on the drawbar but also for three point linkage equipment. Built to the dimensions of British Standard, Category 2, the linkage is actuated by a hydraulic cylinder and rockshaft mounted under the rear frame cover. Hydraulic power is supplied by the hydraulic Lift-All pump. The stable heavy-duty draft links have pin height adjustment of 1 ¾" to 36 ¼" above ground level. Simple, effortless adjustments on the linkage arms provide as great a variety to be desired. The two vertical links which connect to the rockshaft to the draft links, may be allowed to float independently of may be locked at the one height for rigid control.

**Farmall Lift-All: For finger touch control of mounted equipment**

Lift-All cylinders provide three types of hydraulic lift of mounted equipment – regular lift, delayed lift or selective lift.

1) Regular lift raises and lowers all sections of the implement simultaneously
2) Delayed lift raises the rear section after the front, so that all sections enter of leave the ground at the one point, and provides thorough cultivation at all times.
3) Selective lift gives independent control of right or left sections when working fields of irregular shape.

**Two-way Hydraulic Remote Control: for finger touch control of trailing implements**

In addition to the three-point linkage, the Super AM is designed for two-way hydraulic Remote Control of trailing implements. The hydraulic remote control attachment is available with the tractor, or can be supplied for Super AM tractors in the field without alteration. With two-way hydraulic remote control the muscle-tiring stretching for out-of-the-way adjusting levers is taken out of the daily work. The remote control cylinder provides power to make on the move working adjustments in fractions of an inch, or to fully raise the implement for transport. This means the work is done faster and better, to save effort and cut costs.

**Specifications**

| | |
|---|---|
| Horsepower – belt, max. | 41 |
| Horsepower – drawbar, max. | 36.7 |
| Engine | 4-cyl |
| Displacement | 264cu. in. |
| Rated Engine Speed | 1,450rpm at full load |
| Bore and Stroke | 4"x 5 ¼" |
| Pistons | low expansion aluminum alloy |
| Piston Rings | 3 compression, 1 oil |

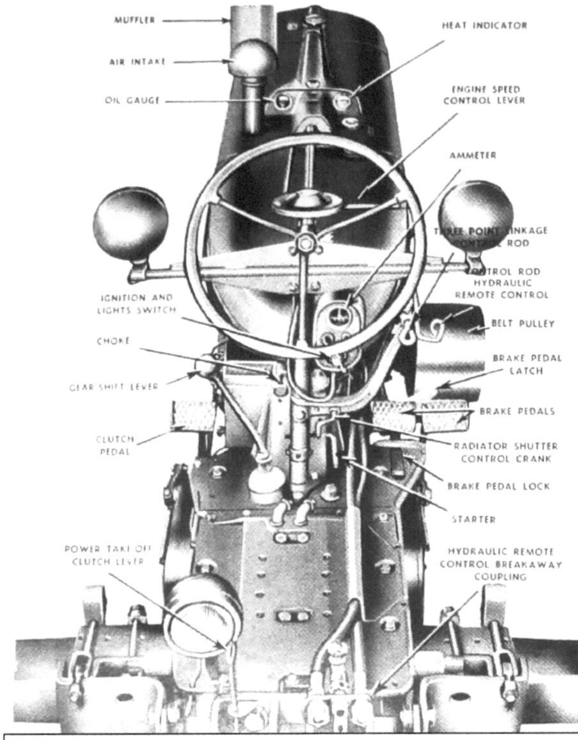


*Well positioned controls give the driver easy and complete control of the Super AM's operation.*

**Electrical Equipment**

Electric Starting, 6-volt battery, automatic spark advance

| | |
|---|---|
| Engine Lubrication | Full Pressure |
| Cooling System | 4 rows tubes, pressure |
| Cooling System capacity | 5 gallons |
| Fuel tank capacity | 17 ½gal kerosene, |
| Starting tank | ¾ gal petrol |
| Belt Pulley – diameter and face | 11" x 7 ½" |
| Belt Pulley – Speed | 899 rpm, 2,588' per min |
| Power Take Off | 537 rpm |
| Steering | Enclosed worm and gear drive |
| Clutch | 11" single plate, Heavy Duty springs |
| Tyres – front | 6.00 x 16 |
| Tyres – rear | 12 x 38 wide base |
| Forward speeds 12x 38 tyres: | mph.1 ¾, 3 ¾, 5, 6,16 ⅝ |
| Reverse speed, 12x38 tyres, mph | 2 ⅜ |
| Front wheel tread (centre to centre) | 8 ½" |
| Rear wheel tread (centre to centre) | 52" – 88" (or 52" – 100", special rear axle) |
| Length | 134" |
| Wheelbase | 88 ½" |

| | |
|---|---|
| Width overall | 84 ½" |
| Width overall | 96 ½" |
| Height to top of radiator | 64 ½" |
| Height to top of steering wheel | 77 ½" |
| Turning Radius, with braking | 8' 3" |
| Turning radius, without braking | 8'6" |
| Clearance under rear axle housing | 26 ½" |
| Clearance under rear frame | 18 ⅞" |

Drawbar vertical adjustments:

13 ⅛", 14 ½",16", 17 ⅝", 19"

Drawbar Lateral Adjustments: 9 ⅛" from centre

Approximate operating weight with rubber tyres, including one set of weights 7,462 lbs

**Standard Equipment**

AC-264 four cylinder, 4" x 5 ¼" engine with aluminum alloy pistons; variable speed governor controlled from seat; manifold heat control valve; replaceable cylinder sleeves; molybdenum-iron alloy exhaust valve seat inserts; 6-volt electrical system with voltage regulation; battery ignition; electric start; oil-bath air cleaner; fuel strainer; floating oil screen; replaceable element oil filter; spark arrestor and exhaust extension; adjustable radiator shutter and heat indicator; enclosed external contracting, steering foot brakes, operation on the differential shaft; spring loaded foot clutch; vertically adjustable rear drawbar; rear swinging drawbar; adjustable, upholstered, waterproof seat; 6.00x 16 6-ply front tyres; 12 x 38, wide base, 6-ply rear tyres on demountable steel rims; 2 front wheel weights; 2 rear wheel weights; 88" tread rear axles.

**Special Equipment**

Belt pulley attachment; rear power take off attachment; electric lighting – two front and one rear; front wheel weights; rear wheel weights, additional if required; exhaust muffler; hydraulic Lift-All pump; three point linkage – British Standard, Category 2; two-way hydraulic Remote-Control, A.S.A.E. – S.A.E. dimensions; 100" tread rear axles.

**Tractor Build Data SAM**

**Financial Year (July 1—June 30)**

| **1953-1959** | **265 Units Built** |
|---|---|
| 1953/1954 | 130 |
| 1954/1955 | 82 |
| 1955/1956 | 51 |
| 1956/1957 | 1 |
| 1957/1958 | 0 |
| 1958/1959 | 1 |

Exact serial numbers for the SAM are unknown at this stage.

## Super AM-D

**Diesel Power** An increasing demand for a diesel -powered Farmall tractor has resulted in the development and the manufacture in Australia by International Harvester of a diesel tractor, the Farmall Super AMD. Released 18 December, 1953, its 50 hp engine (the AD-264) provides ample power for most farm requirements. The Farmall Super AMD not only uses cheap diesel fuel but for greater economy meters the fuel that is used.

Its power is well balanced with the selected field speeds and the increased traction provided by the new, wide based tyres. The lift-all cylinders, three point linkage and two way hydraulic remote control give the tractor driver full control from the tractor driver full control from the tractor seat of both mounted and trailing implements.

Adjustable rear wheel tread and high clearance beneath the frame enable the tractor to "fit" any crop, whether it be on broad acres, mixed farming or row crops of maize and sugar cane. The position of seat and controls assures comfort and convenience of operation. The operator sits on a comfortable hydraulically damped seat, well up out of the dust zone and with uninterrupted vision of all working points.

**Clearance, Convenience and Adaptability**
The 26 ½" axle housing clearance of the Super AMD makes this the ideal tractor for row crop cultivation. This clearance enables the farmer to cultivate tall row crops without fear of damage to the crop.

**Positive Traction** new tyre sizes assure positive traction, to use the extra power in the Super AMD engine for heavy drawbar work. The Super AMD has an extremely wide range of rear wheel tread adjustment, combined with high under axle clearance for cultivation all kinds of row crops.

- **Australian Diesel** 50 hp, AD264 diesel engine, made in Australia
- **Clearance** Extra high clearance under the rear axle housings for row-crop cultivation
- **Adjustable Rear Tread** The rear wheels are adjustable from 52" – 100". The Super AMD will "fit" any crop
- **Quick, All-Diesel Starting** Glow Plugs and heavy duty, solenoid-operated starter motor assure easy starting under all conditions.
- **Complete Combustion** Swirl-type pre-combustion chambers assure complete fuel combustion and minimize "diesel knock"
- **Economical Operation** Design of engine and fuel injection equipment provides efficient use of fuel
- **Balance** Balanced tractor power – weight distribution balanced with traction and stability, to provide smooth operation
- **Turning Radius** The Super AMD will turn in less than its own length, headlands are reduced to a minimum. Maximum time is saved in turning
- **Low Maintenance** Low injection pressure with single orifice injection minimizes fuel pump and injector maintenance and provides longer life.
- **Sensitive Governing** Sensitive, full-speed range governor maintains constant engine speeds under varying loads
- **Lugging Power** An exclusive torque control device increases the ability of the engine to "hang-on" without stalling under extreme loads
- **Versatility** Three-point linkage, hydraulic Remote-Control and lift-all cylinders make the Super AMD adaptable to all types of Australian agriculture

**Easy to Maneuver in Row Crops** The Farmall Super AMD has only an 8'3" turning circle with steering brakes applied and 8'6" without steering brakes – Less than its own length! This great maneuverability is indispensable to row crop farmers for short turning. Steering wheel position and steering gear mechanism give easy, ready response and result in shorter headlands and maximum time saved on any farm.

**Full Hydraulic System** The Super AMD is designed with a full hydraulic system. Pressure supplied by the famous Lift-All pump, gives convenient and positive control of direct mounted equipment, three-point linkage equipment and Remote-Control of trailing equipment.

**Efficient Diesel Engine** The AD-264 is a four cylinder, vertical, overhead valve diesel engine with replaceable cylinder sleeves, 4" bore and 5 ¼" stroke. Its piston displacement is 264 cu. in. The fuel injection system employs a CAV pump with a sensitive centrifugal governor and exclusive torque control device. Low fuel-injection pressure is made possible with the swirl type pre-combustion chamber, which assures greater power and economy from complete fuel combustion, provided by the maximum intermixing of the air and injected fuel. Equipped with glow plugs and a solenoid-operated starter motor, the engine is quickly started on diesel fuel.

**Hydraulic Three Point Linkage: For finger touch control of direct connected equipment**

The Super AMD is designed not only for the hard, tough pulling work on the drawbar but also for three point linkage equipment. Built to the dimensions of British Standard, Category 2, the linkage is actuated by a hydraulic cylinder and rockshaft mounted under the rear frame cover. Hydraulic power is supplied by the hydraulic Lift-All pump. The stable heavy-duty draft links have pin height adjustment of 1 ¾" to 36 ¼" above ground level. Simple, effortless adjustments on the linkage arms provide as great a variety to be desired. The two vertical links which connect to the rockshaft to the draft links, may be allowed to float independently of may be locked at the one height for rigid control.

**Standard Equipment**

AD-264 four cylinder diesel engine, 4" x 5 ¼" stroke. Diesel injection pump with four 8mm plungers, manual primer pump, excess fuel device, centrifugal governor, automatic torque control and single orifice injectors; glow plugs incorporated in nozzle bodies; two 6-volt, 17 plate batteries; voltage regulator; electric starting with solenoid operated heavy duty starter motor; replaceable cylinder sleeves; 7" oil-bath air cleaner; replaceable cartridge type fuel filter; water trap fuel filter; floating oil screen; replaceable element oil filter; spark arrestor, exhaust extension; heat indicator; oil pressure indicator; enclosed external contracting, steering foot brakes, operation on the differential shaft; spring loaded foot clutch; vertically adjustable rear drawbar; rear swinging drawbar; adjustable, upholstered, waterproof seat; 6.00x 16 front tyres on demountable rims; 12 x 38, wide base, rear tyres on demountable steel rims; 2 front wheel weights; 2 rear wheel weights.

**Optional Equipment**

11" x 7 ½" Belt pulley attachment; rear power take off attachment; electric lighting – two front and one rear; ASAE-SAE standard rear power take-off; hydraulic pump unit; three point linkage – British Standard, Category 2; two-way hydraulic Remote-Control unit; lift all pump; rear wheel weights; extra front wheel weights, 100" axles.

**Specifications**

| | |
|---|---|
| Horsepower – belt, max. | 46.5 |
| Horsepower – drawbar, max. | 42 |
| Engine | 4-cyl |
| Displacement | 264cu. in. |
| Rated Engine Speed | 1,450rpm at full load |
| Bore and Stroke | 4"x 5 ¼" |
| Pistons | low expansion aluminum alloy |
| Piston Rings | 3 compression, 2 oil |
| Injection | CAV pump with four 8mm plungers |
| Injection Nozzles | Single orifice injecting into pre-combustion chamber |
| Injection Pressure | 700 to 710 lb. per sq.in. |
| Starting | Direct on Diesel fuel |
| Starting System | Glow plugs, operated by two position switch in combination with starter switch |
| Compression Ratio | 16.5 to 1 |
| Electrical Equipment | Two 6 volt 17-plate batteries connected in series |

| | |
|---|---|
| Engine Lubrication: | Full Pressure |
| Cooling System | 4 rows tubes, pressure |
| Belt Pulley – diameter and face | 11" x 7 ½" |
| Belt Pulley – Speed | 899 rpm, 2,588' per min |
| Power Take Off | 537 rpm |
| Steering | Enclosed worm and gear drive |
| Clutch | 12" single plate |

| | |
|---|---|
| Tyres – front | 6.00 x 16 |
| Tyres – rear | 12 x 38 wide base |
| Forward speeds mph | 1 ¾, 3 ¾, 5, 6, 16 ⅝ |
| Reverse speed, mph | 2 ⅜ |
| Front wheel tread | 8 ½" |
| Rear wheel tread | 52" – 88" (or 52" – 100", special rear axle) |
| Length | 134" |
| Wheelbase | 88 ½" |
| Width overall (outer edge of 88" rear axle) | 84 ½" |
| Height to top of steering wheel | 77 ½" |
| Turning Radius, with braking | 8' 3" |
| Turning radius, without braking | 8'6" |
| Clearance under rear axle housing | 26 ½" |
| Clearance under rear frame | 18 ⅞" |
| Drawbar height adjustments: | 13 ⅛" to 19" |
| Drawbar Lateral Adjustments | 18 ¼" |

Approx. operating weight w/out operator 6,804lbs

**Tractor Build Data Super AMD**

| **1953-1959** | **472 Units Built** |
|---|---|
| 1953/1954 | 120 |
| 1954/1955 | 200 |
| 1955/1956 | 92 |
| 1956/1957 | 52 |
| 1957/1958 | 1 |
| 1958/1959 | 7 |

Exact serial numbers for the Super AMD are unknown at this stage.

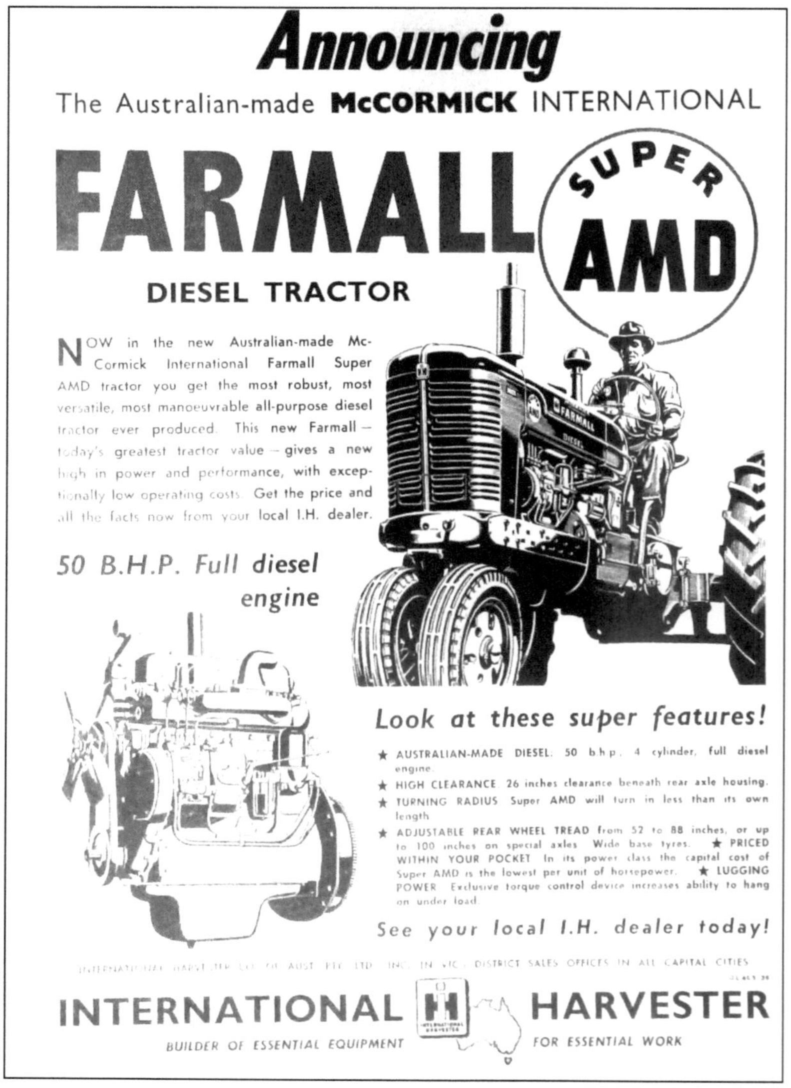

# Chapter 3
# AOS-6

## AOS-6 – Orchard and Standard

Australian Orchard Special (AOS) is a modified W-6 tractor with an AC-248 engine which was introduced 3 June 1953. It was built for specialty farmers needs from 1953 to 1959. Great in apple, citrus and fruit orchards as well as vineyards for its low profile and narrow frame, along with a lowered seat and smaller tyres, the AOS-6 Tractor came as an Orchard, with the Standard version being optional.
Following the success of the demonstrator models in the USA during 1951 as a promotional tool, the first AOS-6 tractors were painted white.
The earliest reference to the white tractors is on 8 September 1953, It is unknown how many tractors were painted this way. It could be anywhere from as few as six (how many were built in the first year) or in the hundreds.

R. Devlin, former Chief Product Engineer for IH, recalls that quite a few of the white AOS-6 tractors came off the line. He wasn't excited about the white paint, as it showed who had grease on their clothing or hands and if anyone had spilled oil on the tractor. He was glad when the salesmen decided to return to the more forgiving red paint.

**Orchard Tractor** the McCormick International AOS-6 orchard tractor with its side outlet exhaust, low overhead clearance, full hydraulic system and wide-base tyres has the tractor power needed for all the jobs in the orchard. The heavy-duty standard 3-point linkage is easily controlled by a small lever close to the steering wheel.
A channel-type toolbar for a full line of toolbar connected implements can be simply attached to the 3-point

linkage to provide hydraulic control and efficient cultivation in orchards and vineyards.

**Standard Tractor** Regularly equipped with British Standard, Category 2, 3-point linkage, the AOS-6 has a wide range of use on any farm. With the 5 forward speeds from 1 ½ mile to 13 ¾ mph, the AOS-6 has the right speed for every job, from the operation of an automatic pick-up baler or ensilage harvester to the fast working speeds for spreading fertilizer and for transport around the farm. The full range of matched 3-point linkage machines means low-cost operation on the dairy farm and a versatile, powerful tractor for the grazier and large, diversified operations.

**Full Hydraulic System** The 3-point linkage of the AOS-6 Orchard Tractor is designed to British Standard, Category 2 dimensions for 3-point connected equipment such as the offset orchard disc harrow. The AOS-6 Orchard Tractor can also be equipped with a remote control attachment with self sealing breakaway couplings and a two-way hydraulic cylinder with ASAE standard dimensions for operating trailing equipment such as the GL-14 disc harrow.

*Above: Left hand view of the AOS-6 tractor showing the downdraft exhaust*

*Right: Right-hand view of the AOS-6 tractor showing the updraft which is supplied as special equipment*

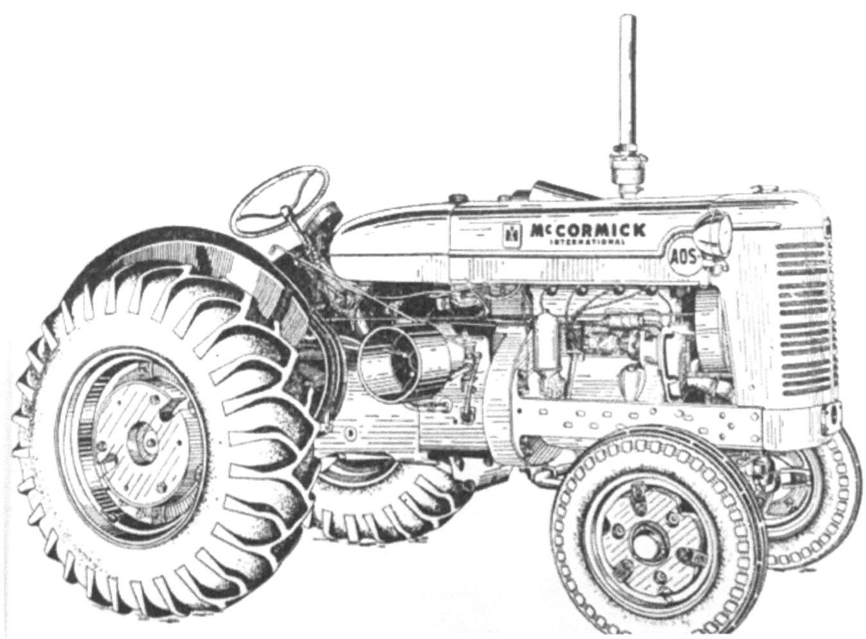

### Standard Equipment

Four-cylinder kerosene engine; battery ignition with automatic advance and retard on distributor; variable speed governor, controlled from seat; oil-type air cleaner; oil filter with replaceable element; floating oil screen; fuel strainer; adjustable radiator shutter and heat indicator; manifold heat control valve; replaceable cylinder sleeves; exhaust valve-seat inserts; enclosed external-contracting steering foot brakes operating on differential shafts; spring-loaded foot clutch; front drawbar; adjustable seat; spark arrestor and vertical exhaust; two 6.00 x 16 6-ply front tyres with demountable rims; two 13 x 26 6-ply wide base rear tyres with demountable rims; six-volt electric starting; rigid vertically adjustable drawbar.

### Specifications

| | |
|---|---|
| Engine | AC-248 |
| Max. belt horsepower | 37 |
| Max drawbar horsepower | 32.5 |
| Variable speed control governed from seat – full load engine speed range, RPM | 950-1450 |
| Forward Speeds, MPH (with 13x26 rear tyres) | 1 ½, 3 ⅛, 4, 5⅝, 13 ¾ |
| Reverse Speed, MPH (with 13x26 rear tyres) | 1⅞ |
| Bore and stroke, inches | 3 ⅞ x 5 ¼ |
| Number of cylinders | 4 |
| Replaceable cylinder sleeves | Yes |
| Belt pulley, diam and face, inches | 11 x 7 ½ |
| Belt pulley, RPM at rated engine speed | 899 |
| Power take-off, RPM | 537 |
| Front wheel tread, inches | 47 |
| Drive wheel tread, inches | 53 |
| Wheelbase, inches | 76 ⅜ |
| Total Length, inches | 117 |
| Total width, inches | 68 ⅞ |
| Height overall, top of steering wheel, | 63⅞ in. |
| Turning radius, feet (without braking) | 9ft. 10in. |
| Cooling system | Pump, thermostat |
| Capacity of cooling system, imp. Gallons | 5 |
| Fuel | Kerosene |
| Capacity, fuel tank, imp. Gallons | 17 ½ |
| Capacity, auxillary tank, imp. Gallons | ¾ |
| Engine lubrication | Pressure |
| Clutch | 11 inch, single plate |
| Steering | Enclosed worm and gear |
| Drawbar adjustment, inches above ground | 9 ½, 11 ¼, 13, 14 ¾, 16 ½ |
| Swinging Drawbar lateral hitch adjustment | 18 ¼ |
| Approximate shipping weight, lbs. | 5,162 |

### Optional Equipment

Engine-mounted Plessey hydraulic pump; hydraulic control; 3-point linkage; belt pulley; rear power take-off; electric lighting; swinging drawbar; 2-way remote control attachment; 2 sets of front and rear wheel weights.

### Tractor Build Data
### (per financial year, July 1-June 30)

1,920 Units Built— Production numbers are not available for the individual tractor variations.

Sold from 1953 to 1959

Cost in 1959 £ 1,143 or $2,286

| Year | Units Built |
|---|---|
| 1952/1953 | 6 |
| 1953/1954 | 868 |
| 1954/1955 | 800 |
| 1955/1956 | 150 |
| 1956/1957 | 75 |
| 1957/1958 | 7 |
| 1958/1959 | 14 |

# Chapter 4
# AW-7

The AW-7 is a product of Australian workmanship, manufactured in International Harvester's farm equipment works in Geelong, Victoria and is another achievement in the proud record of the Australian farm machinery industry.

Introduced in 1957, the McCormick International AW-7 replaces the AW-6 tractor. The official presentation for the AW-7 Diesel was held in Perth, Western Australia on July 20, 1957. The official release of the kerosene version followed in March, 1958.

It is available with either the diesel (50hp) or kerosene (45hp) engine. Most interesting feature about this big all purpose farm tractor is its power steering, which allows the big unit to be handled like a small car. The paint scheme is all red, including tyre rims and weights with white grille, cast aluminum badge, and white background for bonnet side emblems

In demonstrations throughout the country, farmers have been pleasantly surprised how the tractor performs when driven across rough ground without movement of the front wheels and without deviation from its course.

One farmer said, "I don't have to fight the wheel and I still feel fresh at the end of a long day's driving."

The AW-7 is a big all-purpose tractor. Its power is balanced with well-selected field speeds and the traction afforded by wide-base tyres. Power steering simplifies tractor operation by reducing turning effort. Comfort and ease of handling are completely assured by the newly positioned seat and controls. Live hydraulics ensures continuous hydraulic operation. Three point linkage and two way hydraulic remote control gives the tractor driver all around control from the tractor seat of both direct connected and trailed implements.

**Standard Equipment**: AW-7 features an Australian built AC-264 four cylinder engine in either kerosene or diesel fuels; variable speed governor controlled from seat; manifold heat control valve; replaceable cylinder sleeves; 6-volt battery with voltage regulation; battery ignition; manually controlled starter motor; oil-bath type air cleaner; fuel strainer; floating oil screen; replaceable element oil filter; spark arrester and vertical exhaust; enclosed external -contracting steering foot brakes, operation on the differential shafts; spring loaded foot clutch; front drawbar; adjustable rear drawbar; swinging rear drawbar; adjustable, upholstered waterproof seat; 7.50 x 16 6-ply front tires on demountable rims; 14-30 6 ply wide base rear tires on demountable rims; 2 front wheel weights; 2 rear wheel weights; standing platform; power steering; toolbox.

**Optional Equipment:** Electric lighting – two front and one rear lamp; 11in x 7 ½ in belt pulley attachment; rear power take-off; hydraulic pump unit; three point linkage – British Standard, Category 2; hydraulic Remote-Control unit; rear wheel weights; 15-30 wide base tires; knapsack spray carrier; detachable seat pad.

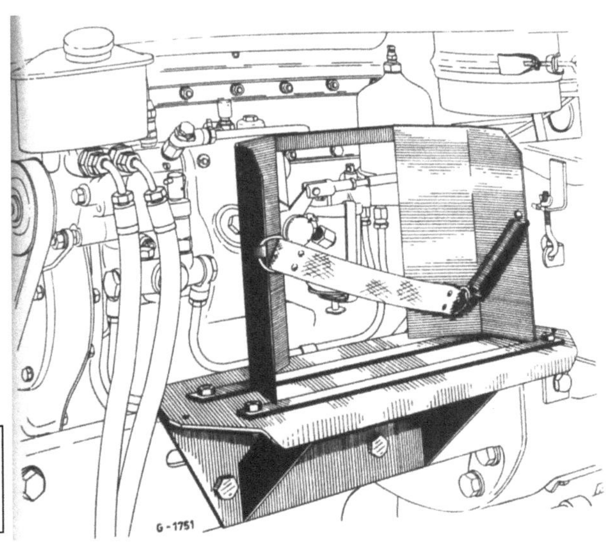


*Knapsack Spray Carrier is attached to the tractor behind the front frame channel.*

*Right: A trainload of 18 AW-7 tractors recently dispatched to IH dealers leaves the railway marshalling yards at Fremantle, Western Australia for country centers.*

**Matched Equipment:**

**Pusher Plate** – This pusher is sturdily bolted and braced to the main frame, just below the engine compartment and will stand up to the rugged conditions of pusher-plate operation.

**Power Digger** – the International Power Digger has a particularly wide range of operation, but is primarily used as a trench digging machine. The 3.5 cu ft capacity bucket lifts to a maximum height of 11 feet, digs to a depth of 8 feet, has a maximum reach of 10'6" and has a 140° slewing arc. Operator has an unobstructed view of every movement of the bucket and the work being undertaken; all controls are at the operators fingertips.

**Engine Specifications:**

**Engine, Kerosene:** Four cylinder, vertical overhead valve type. Bore 4 in., stroke 5 ¼ in., displacement 264 cu. in. Engine speed at rated load 1450 r.p.m. Cylinder sleeves; dry-type, replaceable, fine-grain centrifugal castings. Low expansion, aluminum alloy pistons; three compression rings each piston, bottom ring taper-faced, one oil ring each piston. "Lucas" battery ignition with automatic spark advance. 1 ¼" updraft carburetor – centrifugal, mechanical, variable speed type governor with manual control. Full pressure lubrication with removable oil -filter element. Pressure type cooling system, radiator has 4 rows of tubes, centrifugal pump and thermostat; temperature controlled by radiator shutter. Compression ratio 5.07:1. Cooling system capacity 5 gallons, fuel tank capacity 17½ gallons, auxiliary tank ¾ gallons. Electrical system – one 6-volt, 17 plate battery, "Lucas" starting motor; belt driven generator with voltage regulation.

*Photo Right: AW-7 with Power Digger attached.*

*Mr. Jack Eccles (far left) -equipment engineer for Auckland City Council, New Zealand, paid a visit with his wife to IH's Geelong works.*

**Engine, Diesel:** Four cylinder, vertical overhead valve type. Bore 4in, stroke 5 ¼ in., displacement 264 cu. in. Engine speed at rated load 1450 r.p.m. Cylinder sleeves; dry type, replaceable, fine grain centrifugal castings. Low expansion, cam-ground, aluminum alloy pistons; three compression rings each piston, top ring chromium-plated and taper-faced, bottom ring taper-faced; two oil ring chromium-plated and taper faced, bottom ring taper-faced; two oil rings each piston, one above, one below piston-pin. C.A.V. injection pump with four 8mm plungers, injection pressure 1544 to 1617 psi; centrifugal mechanical governor incorporated in fuel pump, operative at all speeds. Direct starting on diesel fuel, starting aid – glow plug nozzle bodies operated by two-position switch in combination with starter switch. Full pressure lubrication with removable oil filter element. Pressure type cooling system, radiator has four rows of tubes; centrifugal pump and thermostat. Compression ratio 16.5:1. Cooling system capacity 5 gallons. Fuel tank capacity 17½ gallons. Electrical system – 12 volt – two 6-volt 17 plate batteries connected in series, "Lucas" positive engagement, manually operated 4 7/16 " diameter, starting motor; belt driven generator with voltage regulation.

*Below: International Harvester, first to build tractors in Australia, was one of only two manufacturers in 1958. Here, AW-7 tractors are coming off the Geelong Assembly Line.*

**Specifications:**

| Horsepower: | Kerosene | Diesel |
|---|---|---|
| Max belt | 41 | 48.8 |
| Max Drawbar | 36.7 | 42 |

| Clutch: | Kerosene | 11in |
|---|---|---|
| | Diesel | 12in single plate |

**Steering:** Enclosed worm and gear type

**Brakes:** External contracting bands on cast iron drums, foot operated, singly or interlocked on both wheels.

**Belt Pulley**

Diameter and face width 11in x 7½ inches

Speed 899 rpm – 2588 fpm

Power Take Off: 537 rpm

**Tyres**

Front: 7.50 -16, 6-ply on demountable rims

Rear: 14-30, 8 ply on demountable rims

**Drawbar Height adjustments**:

12 ¾ in., 14 ¾ in., 16 ¾ in., 18 ¾ in., 20 ¾ in

Lateral adjustments 9 1/8 in. each side of tractor centre line.

**Road Speeds, mph:** developed at 1450 rpm.

| Forward | 2.460, 3.508, 4.560, 5.520, 15.420 |
|---|---|
| Reverse | 3.295 |

**General Dimensions: inches,**

| | |
|---|---|
| Turning radius with brakes applied | 132 |
| Without brakes | 156 |
| Length overall (inches) | 125 |
| **Width overall** | |
| (to outside edge of rear tire) | 72 ⅜ |
| Height (to top of pre-cleaner) | 73 ¾ |
| Wheelbase | 76 ½ |
| Front Wheel tread | 52 ¼ |
| Rear wheel tread | 57 |
| Clearance under rear axle | 15 ⅛ |

| **Weights:** | **Kerosene** | **Diesel** |
|---|---|---|
| Shipping | 5538lb | 5768lb |
| Operating | 7068lb | 7298lb |

*Some of the 'back-room' boys inspecting the AW-7 motor on the dynamometer.*

*A group of Western Australian IH dealers gather around the recently announced model AW-7 McCormick International tractor, introduced by John Davies*

**Tractor Build data**

1956-1961

2408 Units Built

Cost in 1959

| | |
|---|---|
| Diesel | £1,575 ($3,150) |
| Kerosene | £1,400 ($2,800) |

Australian Tractor Test No. 30 – International AW-7 Diesel

**Diesel**

| | |
|---|---|
| 1956/1957 | 106 |
| 1957/1958 | 871 |
| 1958/1959 | 1244 |
| 1959/1960 | Unknown |
| 1960/1961 | Unknown |

**Kerosene**

| | |
|---|---|
| 1956/1957 | 1 |
| 1957/1958 | 46 |
| 1958/1959 | 140 |
| 1959/1960 | Unknown |
| 1960/1961 | Unknown |

*Above: Rear view of AW-7 Kerosene tractor*

*Left: AW-7 tractor being loaded onto rail cars for shipment around Australia.*

*Below Right: Line drawing of AW-7*

# Pre-Production Photos of the Diesel AW-7

Most noticeable on this tractor is the red grille, second toolbox, and the cast badge that reads AWD-7

Photos are taken outside of Geelong Works and are dated 23 January 1957

*Above: Advertisement for the "New" AW-7 Diesel Tractor*

*Left: An International truck and McCormick International AW-7 Diesel tractor for presentation under the Colombo Plan to the Chulalong Korn University in Bangkok, Thailand. June, 1958*

# Chapter 5
# AM-7

Replacing the Farmall AM and the Farmall AMD are the AM-7K and AM-7D Farmall tractors. Introduced with little fanfare, this tractor was simply an updated version of the AM and AMD tractors. The AM-7 Diesel was introduced in July of 1957 while the AM-7 Kerosene was released later on the 28 March, 1958.

The AM-7K still used the AC-264 motor, and the AM-7D had the same AD-264 motor as well. The paint scheme changes slightly, with a white grille and white background on the cast aluminum badges.

Fully adjustable wheel treads from 52-100 inches, 26 ½ inch rear axle clearance, and turning circle less than its own length and a 50hp (brake) diesel engine; 45 brake hp kerosene motor matches the AM-7 to the big row crop farms.

The Farmall tractor is unmatched in uninterrupted vision over the crop and can be equipped with a full line of mounted row crop cultivation equipment.

**Specifications**

**Engine**

| | |
|---|---|
| Cylinders | 4 |
| Bore and Stroke | 4 x 5 ¼ inches |
| Engine speed, rpm | 1450 |
| **Clutch** (spring loaded type) | 11" |

**Belt Pulley**

| | |
|---|---|
| Belt pulley speed | 899rpm |
| Belt speed (feet per minute) | 2588 |
| Diameter and Face | 11" x 7 ½" |
| PTO Speed | 537 rpm |
| Foot Brakes | External contracting on drums |

**Speed** on 12x38 tyres

| | | | |
|---|---|---|---|
| 1st | 1.77 | 2nd | 3.81 |
| 3rd | 4.95 | 4th | 6.00 |
| 5th | 16.70 | Reverse | 2.32 |

**Wheels and Tread**

| | |
|---|---|
| Front wheels | 6.00 x 16 |
| Rear Wheels | 12 x 38 |
| Tread – front | 8 ½" |
| Tread – rear with 88" rear axles | 52-88" |
| Tread – rear with 100" rear axles | 52-100" |
| Wheelbase | 88 ½" |

**General Dimensions**

| | |
|---|---|
| Length overall | 134" |
| Width overall (88" rear axles) | 84 ½" |
| Width overall (100" rear axles) | 96 ½" |
| Height overall (to top of radiator) | 64 ½" |
| Height overall (to top of steering wheel) | 78" |
| Clearance under rear axle | 27" |

Rear view of the AM-7 Diesel tractor

| | |
|---|---|
| Clearance under rear frame | 18 ½" |
| Drawbar (lateral hitch adjustment) | 18 ¼" |
| Drawbar adjustment from ground | 13 1/8", 14 ½", 16", 17 5/8", 19" |
| Turning radius (with brake applied) | 8' 3" |
| Turning radius (without brake applied) | 8'6" |
| Fuel tank (Kerosene or Diesel) | 17 ½ gallons |
| Petrol tank—for starting only | ¾ gallon |
| Water cooling system | 5 gallons |
| Crankcase Pan | 6.4 quarts |
| Transmission and differential case | 10.4 gallons |
| Air cleaner oil cup | 1.8 pint |
| Steering gear housing | 1.2 pint |

**Weight**

| | |
|---|---|
| Approximate shipping weight | 5,650 lbs |
| Operating weight | 7, 016 lbs |

**Tractor Build Data**

Built from 1957-1961

15 AM7-K Units Built

143 AM7-D Units Built

| | Kerosene | Diesel |
|---|---|---|
| 1957/1958 | 8 | 78 |
| 1958/1959 | 7 | 65 |
| 1959/1960 | Unknown | |
| 1960/1961 | Unknown | |

| | |
|---|---|
| Kerosene price in 1959 | £ 1,300 ($2,600) |
| Diesel price in 1959 | £ 1,475 ($2,950) |

**Kerosene Motor on AM-7;** Only 15 tractors were made with this engine.
Date on tractor photograph is 18 September 1958

# Chapter 6
# A-554

The International A-554 was styled from the AW-7, with modernized looks and some mechanical upgrades. This gave the tractor a true Australian wheatland look. The International A-554 was introduced in 1961 and features included a choice of horsepower, 55hp diesel engine or a 52hp kerosene engine; power steering; channel steel frame; vertically adjustable rear roller drawbar; optional power take off.

The International A-554 was one of the more popular tractors built. A 1963 Engineering and Development report states the estimated production life would be two years. The tractor proved popular and the A-554 was built for five years. It was offered in diesel and kerosene, with the diesel being standard. The kerosene engine was offered in the first year of manufacture in 1962, as a special order. It is unknown how many kerosene tractors were built as diesel power had taken hold in the farming world, offering more horsepower, better fuel consumption and lower running costs than the kerosene engines. All tractors made in Geelong after 1965 were only available as a diesel engine tractor.

An industrial version was also produced, known as the A-2554. The row crop version, the Farmall A-554, was introduced in early November 1964, replacing the A-514.

According to a 1966 brochure, the tractor was mechanically improved – "11 Ways Better." The eleven ways were:

- New Heavy-duty Dyna-Life clutch
- More durable and increased capacity transmission
- New fuel injection system
- New muffler-spark arrester combination
- New 18 inch front tyres
- New heavy-duty battery
- Slo-Roto inlet valves – increases valve life, higher engine compression ratio and improved starting
- Long life molded cog fan and power steering belts
- New heavy duty design bull pinion and bull gear

- Tractor meter to indicate ground speeds in various gears, as well as hours of operation
- Power steering available for the Farmall A-554

***Prototype of the A-554, the A-450 and A-504***

*Both of these pre-production photos are nearly identical with the exception of the name of the tractor. There is a temporary decal of the model number on side of the A-450, while the A-504 has a cast aluminum badge. The A-554 sported a cast aluminum badge in final production.*

**A-554 Special**

Engineers were interested in an A-554 Special tractor from the beginning of this model development. Features such as a front mount step, wide American-style fenders and a streamlined styling are all features of this concept.

The top photo is dated 31 August 1962; Middle photo dates 28 November 1963 and the Bottom photo is from 17 June 1964

Notice the top two photos have similar bonnet and side panels while the bottom photo has a unique instrument cowling. This tractor was never produced.

**Specifications: A-554 Regular**

**Horsepower**

Brake: 56.9

Max. Belt: 51.7

Max. Drawbar: 50.8

**Engine:**

No. of cylinders: 4

Bore and stroke: 4x5 ¼

Displacement: 264

Speed at rated engine RPM: 1800

Clutch size, inches: 12

Cooling system type: Pump and pressure

Lubrication type: Full Pressure

**Belt Pulley**

Diameter and face, in. 11x7 ½

Belt speed at rated engine speed 3,212

| Speeds, MPH | 16.9 x 30 rear tyres | 18.4 x 30 rear tyres |
|---|---|---|
| 1st gear | 2.4 | 2.6 |
| 2nd gear | 3.5 | 3.7 |
| 3rd gear | 4.5 | 4.8 |
| 4th gear | 5.4 | 5.8 |
| 5th gear | 15.2 | 16.0 |
| Reverse | 3.2 | 3.4 |

**Independent PTO**

Speed at rated engine rpm 543

Spline shaft diameter, in. 1 ⅜

**Tyre Sizes**

| | |
|---|---|
| Front | 7.50 x 18 |
| Rear | 16.9 x 30 |

**Wheel treads, inches**

| | |
|---|---|
| Front | 52 ⅝ |
| Rear | 63 |
| Wheelbase, in | 76 ½ |

**Turning Radius, inches**

| | |
|---|---|
| With brakes applied | 132 |
| Without brakes applied | 156 |

**Minimum Ground clearance, in**

| | |
|---|---|
| Under front axle | 14 ¼ |
| Length – overall | 125 |

**Width – overall, in**

| | |
|---|---|
| On 16.9 x 30 tyres | 80 |
| On 18.4 x 30 tyres | 82 ½ |

**Height** to top of radiator 60 ½

**Weight**

with fuel, operator and water ballast 8,160lbs

**Capacities**

Approximate Imperial Measures

| | |
|---|---|
| Fuel tank | 17 ½ gal |
| Water Cooling system | 5 gal |
| Crankcase pan | 13 pints |
| Transmission case | 10 ¾ gal |
| Air cleaner oil cup | 2 pints |
| Steering gear housing | 1 ¼ pints |

*A-2554: Industrial version of the A-554, comes with heavy-duty front axle. Shown here with front end loader; other optional equipment is a pusher plate attachment, power digger and trailer hitch.*

**Standard Equipment; A-554 Regular**

Diesel engine; 12 volt electrical system; tractor meter; voltage regulator; heavy-duty co-axial starter motor; oil bath air cleaner; floating oil screen; replaceable element oil filter; muffler spark arrester; enclosed external contracting steering foot brakes operating on differential shaft; spring loaded foot clutch; engine driven hydraulic system with double spool valve; vertically adjustable rear roller drawbar; deluxe seat with tilt-back device; tool box; pre screener; pre cleaner; independent power take off; 7.50 x 18 front tyres; 16.9 x 30 rear tyres; front wheel weights (set of 2); rear wheel weights (set of 2)

Optional Equipment; A-554 Regular

Belt pulley; lighting (2 front and 1 rear lamp); CAT II 3-point linkage; hydraulic remote control unit; narrow tread rear axle attachment; second rear wheel weights (set of 2); 16.9 x 30 rear tyres; 18.4 x 30 tyres.

*View of the operators' platform.*

The Farmall A-554 is the row crop version, which was introduced in 1964, replacing the A-514.There were very little changes to the styling and operation of the tractor. Both the A-554 Regular and A-554 Farmall were offered until 1966.

**Specifications: A-554 Farmall**

| | |
|---|---|
| **Horsepower:** Diesel Brake | 56.9 |
| Max. Belt | 51.7 |

**Engine:**

| | |
|---|---|
| Number of cylinders | 4 |
| Bore and Stroke | 4 x 5.25 |
| Piston Displacement | 264 |
| Speed at rated engine RPM | 1800 |
| Clutch size in. | 12 |
| Cooling system type | Pump and pressure |
| Lubrication type | Full Pressure |

**Belt Pulley**

| | |
|---|---|
| Diameter and face, in. | 11x7 ½ |
| Belt speed at rated engine speed | 3,212 |

| **Speeds,** MPH | 13.6 x 38 rear tyres | 14.9 x 38 rear tyres |
|---|---|---|
| 1st gear | 2.6 | 2.7 |
| 2nd gear | 3.8 | 3.9 |
| 3rd gear | 4.9 | 5.0 |
| 4th gear | 5.9 | 6.1 |
| 5th gear | 16.6 | 17.1 |
| Reverse | 3.5 | 3.6 |

**Independent PTO**

Speed at rated engine rpm 543

Spline shaft diameter, in. 1 ⅜

**Tyre Sizes**

Front 6.00 x 16

Rear 14.9 x 38

**Wheel treads, inches**

Front 8 ½

Rear 52 to 100

Wheelbase, in 76 ½

**Turning Radius, inches**

With brakes applied 99

Without brakes applied 102

**Ground clearance, inches**

Under rear axle 26

Under rear frame 17 ½

Length – overall 135

Width – outer edge 96 ½

Height to top of steering wheel 81

**Swinging Drawbar Lateral Adjustment, inches** (each side of centre) 12 ⅝

Swinging Drawbar Adjustment, inches (from ground) 12, 13, 14 ⅛, 15 ½, 16 ⅜, 18⅝

**Capacities** Approximate Imperial Measures

| | |
|---|---|
| Fuel tank | 17 ½ gal |
| Water Cooling system | 5 gal |
| Crankcase pan | 13 pints |
| Transmission case | 10 ¾ gal |
| Air cleaner oil cup | 2 pints |
| Steering gear housing | 1 ¼ pints |

**Standard Equipment; A-554 Farmall**

Diesel engine; 12 volt electrical system; tractor meter; voltage regulator; heavy-duty co-axial starter motor; oil bath air cleaner; floating oil screen; replaceable element oil filter; muffler spark arrester; enclosed external contracting steering foot brakes operating on differential shaft; spring loaded foot clutch; engine driven hydraulic system with double spool valve; vertically adjustable rear roller drawbar; deluxe seat with tilt-back device; tool box; pre screener; pre cleaner; standard CAT II 3-point linkage; 6.00 x 16 front tyres; 14.9 x 38 wide base 6-ply rear tyres on demountable steel rims.

**Optional Equipment; A-554 Farmall**

Belt Pulley; independent power take-off; electric lighting, two front and one rear; additional wheel weights; two-way hydraulic remote control; 106" tread rear wheel attachment, deduct for 13.6 x 38 tyres. Note: for specialized requirements an adjustable wide front axle or single front wheel can be fitted.

*A-554 Standard at work*

**Tractor Build Data**

June 1961- July 1966

8,803 Units Built – Production numbers are not available for the individual tractor variations. (Standard or Rowcrop; Diesel or Kerosene) The Kerosene motor was built up to 1963. In the 1964 IH Product Catalogue it simply states that 52 hp. kerosene model available.

*Above: Regional Inspector General, Pacific Area, for IH Export Company, USA, Mr. P. Guilfoyle seen viewing Australian made McCormick International A-554 tractor in Melbourne with Australia's IH Export sales manager Mr. C.J.E. Brennan, Mr. Guilfoyle was on vacation leave (International Harvester in Australia, 1963)*

Cost in 1962 Diesel - £1,525 ($3,050)
Kerosene - £ 1,375 ($2,750)

Australian Tractor Test No. 36, December 1961

For monthly production list, see Appendix D.

| Financial Year (July 1- June 30) | Units |
|---|---|
| 1960/1961 | 1 |
| 1961/1962 | 2333 |
| 1962/1963 | 1983 |
| 1963/1964 | 1730 |
| 1964/1965 | 1856 |
| 1965/1966 | 815 |
| 1966/1967 | 85 |

# Chapter 7
# A-514

*Chapter cover: 2009 photo of A-514 s/n 879, pulling three bottom plough at field day trials.*
*Above: Studio Photo taken 11-October-1961*

Little is known about the production of the Farmall A-514. What is known is that the tractor was built to replace the Farmall AM-7, and acted as a 'filler' until the Farmall A-554 was released in 1965. It was a temporary tractor that was converted into a production model. This is evident in the sticker-like number badges, which was common to use during development of the tractors.

A 1963 Product Engineering report states that the estimated production life of A-514 is three years. Power Farming Technical Annual lists them for sale from 1962 until 1964. Product Development photos date the tractor for production from October 1961.
Last photos dated November 1964, from this archive, show the development of the Farmall A-554 with A-514 tractor as a base.

The A-514 50.5 hp engine, provides plenty of pull or push power for rear or forward mounted machines. It has a fully live engine driven hydraulic system, completely independent of any other control on the tractor.

Features include adjustable rear wheel tread from 52 inches to 100 inches, job-matched gear speed range, ample mounting pads for forward or mid-mounted implements, Cat 2 three point linkage and uninterrupted vision over the crop.

The cost cutting diesel engine gives fuel economy, fast easy starts, smooth working overhead valves, and exclusive torque control, built into the governor meters extra fuel when the engine speed is reduced by extreme overload to give additional lugging power.

**Specifications**

| | |
|---|---|
| Horsepower -brake | 50.5 |
| Maximum Belt | 48.8 |
| Maximum Drawbar | 44.6 |
| Engine - | |
| Displacement | 264 |
| Rated speed, rpm | 1450 |
| Bore and Stroke, inches | 4 x 5 ¼ |
| Pistons, low expansion aluminum alloy | |
| Piston Rings, | 3 compression, 2 oil |
| Lubrication, | Full pressure |
| cooling system | Pressure |
| Belt Pulley –Diam. and face, in. | 11 x 7 ½ |
| Speed, rpm | 899 |
| Speed ft/min | 2588 |
| Power Take off, rpm | 537 |
| Steering, | Enclosed worm and gear |
| Clutch | Single plate, 12in |
| Tyres – Front | 6.00 x 16 |
| Rear | 13.6x38 std; 14.9 x38-opt. |
| Forward speeds, mph | 1 ¾, 3 ¾, 5, 6, 16 ⅝ |
| Reverse Speed, mph | 2 ¾ |
| Front Wheel Tread, centre to centre, in. | 8 ½ |
| Rear Wheel Tread, centre to centre, in. | |
| Standard | 52-88 |
| Special | 52-100 |

| | |
|---|---|
| Length, inches | 135 |
| Wheelbase, inches | 88 ½ |
| Width overall – outer edge (88in axles) | 84 ½ |
| outer edge (100in axles) | 96 ½ |
| Height – top of radiator, inches | 64 ½ |
| top of steering wheel, inches | 77 ½ |
| Turning Radius – with brakes, inches | 99 |
| -without brakes, in. | 102 |
| Clearance – under rear axle housing | 26 ½ |
| -under rear frame | 18 ⅞ |

*Photos Top Left: Pre-production photo of the operators platform; Right: Front view of A-514 with tyres set to maximum width.*
*Photo Above: Side view of A-514*

**Standard Equipment** Diesel Engine; 12 volt electrical system; voltage regulator; heavy-duty co-axial starter motor; oil bath air cleaner; floating oil screen; replaceable element oil filter; spark arrester and exhaust extension; enclosed external contracting steering foot brakes operating on differential shaft; spring loaded foot clutch; vertically adjustable rear roller drawbar; deluxe seat with tilt-back device; 6.00 x 16 front tyres; 13.6 x 38 wide base 6 ply rear tyres on demountable steel rims; complete with front and rear wheel weights; 88in tread axle; tool box.

**Optional Equipment** Belt pulley; power take off, electric lighting, two front and one rear; additional wheel weights; exhaust muffler; engine driven hydraulic system; British Standard Category 2 three-point linkage; two way hydraulic remote control; 100in tread rear axles; rear tyres 14.9 x 38. Note – for special requirements an adjustable wide front axle or single front wheel can be fitted.

*Photo, Above: Side view of A-514*
*Photo, Below: Optional Row Guide fitted to A-514*

**Tractor Build Data**

October 1961 – November 1964

Price in 1962 £1,325 ($2,650)

Unknown with the following exceptions:

1962 150 built

Highest serial number Author has seen is 918 (418th one built).

*Demonstration by McCormick International A-514 Farmall tractor at IH field day. With adjustable front wheel axle, the tractor was fitted with 14-foot toolbar with five middle busters for furrowing operation.*

The Australian made Farmall A-514 row crop tractor is ideal for sugar cane and cotton areas. Implements can be forward, mid or rear-mounted. The A-514 is capable of operating the widest range of implements available. In addition to normal ploughs, harrows, toolbars and miscellaneous machines, special row crop equipment for the Farmall A-514 includes:

**A30-2 Toolbar** A basic toolbar of 3 ¾ in. square construction with bolt on toolbars available from the following range: 25in outer toolbar; 48in inner toolbar; 39 in inner toolbar; 10 ft. straight toolbar. A wide range of tines and points plus weeder, furrower, and twin disc attachments are available.

**Planter Units** The McCormick International No. 184 planter units are designed for attachment to 3 ¼" square toolbars. Up to 4 units can be fitted to a 12 ft bar on the GL-230 toolbar.

**Cotton Picker** The McCormick International No. 114A low drum and no. 120A high drum Cotton Pickers are readily mounted on the Farmall A-514.

**Mid-Mounted Cultivator** No.461 four row mid-mounted cultivator features a parallel link design for constant pitch at all working depths, side to side rigidity for close cultivation and finger tip hydraulics for precise control. Three models of McCormick International two-row mid or rear mounted cultivators are available. They are the GL-142 and GL-187 cultivators and a GL-143 grubber. All are designed to specially suit the Farmall series and meant for use in general row crop or cane operations.

*Cotton assessment trials being made as part of Cotton Field Days. McCormick International No. 114A cotton picker powered by Farmall A-514 Diesel tractor.*

# Chapter 8
# A-414

Styled from the British built B-414, which was built from 1961 to 1966. The A-414 offered new styling to the range of Australian built tractors. Introduced in October 1963, the A-414 offered a 40hp diesel engine; gearbox with 8 forward and 2 reverse speeds; IH "Vary-Touch" hydraulics – automatic draft and position control hydraulics; disc brakes; differential lock; a choice of two PTO speeds in three combinations; distinctive styling; amongst other ideas.

Sixteen selling points for the New A-414 were:

1) AD-154 diesel engine – 40 hp with indirect injection – with four cylinders of course!
2) Fuel tank capacity 10 ½ gallons – enough for a full day's work in average conditions
3) Eight forward speeds – plus two reverse. A speed for every job
4) Heavy-duty box section adjustable front axle for maximum strength and carrying capacity
5) Automotive-type steering for precise ease of control.
6) "Vary-Touch" means automatic draft control; automatic response control; automatic position control. You choose the setting – "Vary-Touch" does the rest!
7) Dual category linkage – gives you twice the versatility!
8) Three point linkage drawbar – with two alternative fixtures – conforms to British and American hitch standards
9) Choice of three PTO combinations
10) Differential lock – of proven IH design

*When the company's wharf facilities are full of incoming and outgoing shipments of tractors, parts, implements, iron ore and more, another port is used in the heart of Geelong for the overflow. Above is part of a convoy of 38 McCormick International A-414 wheel tractors on their way to Geelong wharf for outgoing shipment. (International Harvester in Australia, 1965)*

11) New disc brakes – self energizing in either forward or reverse travel.
12) New-style radiator grille – easily removable for servicing of radiator
13) Radiator filler cap access panel – for simple efficient maintenance without lifting the bonnet
14) Spacious foot platform is big, comfortable and accessible – gives ample room for standing
15) Built-in operator comfort – uncluttered layout and grouped controls means less effort, less fatigue.
16) Distinctive two-tone styling – modern as the moment.

**Big Power with Smooth Running Diesel Engine** The AD-154 has four smooth running cylinders to give more power at low, wear saving engine speed. Economy in running costs result from the combustion tight seal provided by five ring pistons – the top compression and top oil control rings being chromed for long engine life.

**A Speed for every job** A-414 has ten matched speeds to meet all field conditions; six very practical speeds, two for roadwork plus two ideal reverse speeds.

With ten well spaced, job matched speeds, the A-414 gives you the speed you want for the job at hand. The A-414 transmission is engineered to incorporate a safety factor well above any load likely to be imposed in even the most arduous field conditions.

**Choose from Three PTO Combinations** The A-414 has a range of PTO speeds. There are two PTO speeds and a two speed combination from which to choose. Power for the constant running PTO comes from the second stage of the big-capacity dual clutch. In this way a gear change is made without interrupting the PTO.

*First Australian made A-414 was put to work in a peach orchard at Templestowe, Victoria*

*The first A-414 off the production line, Monday October 14, 1961.*
*The first A-414 was delivered to Mr. K.T. Peters, grazier of Mangoplah Store Creek, by Orange IH dealer Herb Lapham at Orange National Field Days.*

**Reversible dual category 3-point linkage swinging or linkage drawbar** Reversible Category 1 and 2 three point linkage brings versatility to the A-414 and means it can handle all types of mounted implements. Two alternative lower hitch points enable best performance to be gained under varying ground conditions. Swinging drawbar with ample lateral and vertical adjustment guarantees correct line of draft. The majority of linkage implements can be operated with swinging drawbar in position. Perfect A-414 weight distribution, ⅔ on rear and ⅓ on front leads to peak performance with trailed implements.

**Full range of Matched Equipment** Implements available for use with the A-414 tractor: A1-41 Disc and Mould board ploughs; GL-325 Offset Disc harrow; GL-222 Spring Time Cultivator; A30 -4 Open Field Toolbar; MA Series Harrows – pasture, diamond, stump-jump, driven tine and leveling; GA4-1 Rotary Mower; GL-330 series Scoops; GL-326 'Carry-Lift' platform; GL-226 Rear Blade; GA50-1 Concrete mixer; and many more.

**Specifications**

**Engine** AD-154—4 stroke cycle diesel, valve in head, with electric all-weather glow plug starting

| | |
|---|---|
| Brake horsepower (stripped) | 40 |
| Belt Horsepower | 36 |
| PTO horsepower | 36.7 |
| Bore and stroke | 3 ½ x 4" |
| Piston displacement, cubic inch | 154 |
| Rated governed engine speed, rpm | 2,000 |
| Maximum torque, lb. ft. | 115 |
| RPM at maximum torque | 1,300 |
| Compression ratio | 22:1 |

*Employees and management celebrate the 1,000th A-414 built. (23 April, 1964)*

**Electric starting and lighting** – 12-volt electric starting system which includes manually engaged starter motor, glow plugs, resistance indicator, generator, voltage regulator, one 12-volt battery. 12-volt front and rear lighting sets are available as attachments.

**Engine Clutch** – single dry disc, spring loaded
Diameter 11"
Dual clutch – two plate clutch heavy duty friction linings, and loaded by heavy duty Bellville washers
PTO clutch, diameter 9"

**Transmission** – Eight forward speeds and two reverse speeds selected by a gear lever and a high-low shift lever. Foot operated differential lock.

| | |
|---|---|
| Bevel gears, spiral | |
| Tooth angle | 30° |
| Bearings | Tapered roller |
| Ratio | 4.88:1 |
| Power Take-off, six spline | 1 ⅜" |
| Speed at rated governed engine speed | |
| Single Speed | 545rpm |
| Two Speed | 545 and 745 rpm |

**Tractor Speeds** – computed at rated governed engine speed, MPH 13.6 x 28 tyre size

| | | | |
|---|---|---|---|
| L1 | 1.46 | H1 | 4.26 |
| L2 | 2.50 | H2 | 7.28 |
| L3 | 3.56 | H3 | 10.37 |
| L4 | 5.31 | H4 | 15.49 |

Low Reverse 2.22 High Reverse 6.46

**Brakes** – foot operated, totally enclosed, externally mounted acting on bull pinion shafts, one on each side. Foot pedals latch together for road

work and a hand parking brake is provided.

| | |
|---|---|
| Disc Diameter, in. | 6 ½ |
| Effective friction area, each lining, sq. in. | 23.6 |
| Number of linings, each side | 4 |
| Total effective friction area, sq. in. | 188.8 |

**Hydraulic System –** Independent fully 'live' system with 'Vary-Touch' draft and position control.
Working pressure, approx. lb. per sq. in. 2,200
Cylinder, bore and stroke, in. 3x 5 ⅝
Average lifting force at end of lower links, horizontal position, pounds (lb.) 2,600

**Tractor Dimensions**

| | |
|---|---|
| Length overall, inches (in.) | 114 ⅜ |
| Tread, front, in 4 inch stages | 48 - 76 |
| Tread, rear, in 8 steps | 48 - 76½ |
| Width, overall: | |
| Front wheels, in. | 59 – 87 |
| Rear wheels, in. | 62 –90½ |
| Height to top of steering wheel, in | 61 ½ |
| Wheelbase | 76 ½ |
| Ground Clearance, under rear frame | 17 ¼ |
| **Turning radius**, inches | |
| Without brakes applied | 133 |
| With brakes applied | 114 |

*First shipment of Australian made A-414 tractors to leave Geelong Works.*
*Bound for Orange, New South Wales. October 26, 1963*

**Capacities, imperial measures**

| | |
|---|---|
| Cooling system, gallons | 2 ½ |
| Fuel Tank, gal | 10.6 |
| Engine lubrication, pints | 9 |
| Transmission case, gal | 4 |
| Hydraulic reservoir, gal | 2 ½ |
| Air cleaner cup, pints | 1 |

**Operating Weight, 13.6 x 28 tyres –** Basic tractor, including fuel, oil, coolant, hydraulic system, three-point linkage and constant running PTO  3,890lb.

**Standard Equipment**
4-cylinder, AD-154, 40bhp diesel engine with replaceable cylinders and 12-volt electric starting with manual engagement; Key-type starter cut-out switch; Heat Indicator, tractor meter; Outside hood air intake; Upswept exhaust with spark arrester Independent foot operated steering brakes (disc) with hand parking control; Differential Lock, foot operated; Rear wheel fenders; Adjustable tread front and rear wheels, front wheels 48-76in., rear wheels 48-76 ½ in.; Deluxe seat; Three point linkage with live hydraulics with 'Vary-Touch' draft control, position control, flow rate control and isolating drive; Constant running 545 rpm PTO; Tyres: 6.00 x 16 front, 13.6 x 28 rear.

*Above Left: Standard A-414*
*Above Right and below: A-414 Hi Clear factory photos. Date reads 28- October 1963*

**Optional Equipment**
Two speed 545 and 745 rpm constant running PTO; Single clutch 545 rpm PTO; Belt Pulley; Down swept Exhaust; Front and rear lighting; Three point linkage drawbar; Swinging drawbar and frame; Self Sealing coupling; Front and rear wheel weights; Front end weight; Double acting hydraulic remote control attachment; Single acting Hydraulic remote control attachment; Tyre Equipment: 6.00 x 16 6-ply front; 7.50 x 16 6-ply front; 13.6 x 28 6-ply rear; 14.9 x 28 6-ply rear; 14.9 x 24 6-ply front; 16.9 x 28 6-ply rear

**A-414 Hi-clear**
The A-414 has 26 3/8 inch clearance under the front axle, and a wide range of adjustability in the tread widths. Diff lock is standard equipment, and mounting pads for mid-mounted and front end equipment. Tyre size, Front—6.00x 16; rear—12.4x 38. optional rear tyres; 13.6x 38.

**Tractor Build Data**

October 1963- November 1966

| | |
|---|---|
| Cost in 1964 | £1,095 ($2,190) |
| Hi-Clear | £1,142 ($2,284) |

5,414 Units Built - Production numbers are not available for the individual tractor variations. (Standard or Hi Clear)

For monthly production list, see Appendix D.

| Financial Year (July 1– June30) | Units |
|---|---|
| 1963/1964 | 1500 |
| 1964/1965 | 2321 |
| 1965/1966 | 1161 |
| 1966/1967 | 432 |

**Pictorial timeline — Building the A-414 Tractor.**

*Above:* Pouring Engine Block

*Above:* Drilling Engine Block

*Above:* Drilling Transmission Case

*Above:* Assembling Transmission

*Left:* Milling Transmission Case

Engine Test

Assembly Line—Front

Assembly Line—Rear

*Right:* Assembly Line—Front

Spray Painting

Final inspection

Driving off the
Assembly line

Loading A-414 tractors.

*Sign reads:*
First shipment of Australian made A-414 tractors to leave Geelong Works.
Bound for Orange, New South Wales.
October 26, 1963

Press conference at a Television Studio, Orange, New South Wales

*Left:* Unveiling the A-414 at Orange Field Days, New South Wales, 1963

# Chapter 9
# 564

The McCormick International 564 replaces Australia's most popular Wheatland tractor – the A-554 in 1966. The 564, with its 264 cubic inch displacement engine is the last of the true Australian designed series. All tractors after this model were components—drive trains from America, engines from Germany, with panels, wheels and minor additions being Australian design—much the same way that the AW-6 was designed and sold.

The 564 has the ability to pull the heaviest trailing implements without effort, working deep into hard unwilling soil. Wide strips of ground can be torn up using three-point linkage or trailing tillage equipment: power absorbing PTO machinery can be driven with all the stamina expected of a much larger tractor.

**Forward Entry**
Ease of entry is a feature of the tractor. A rung and solid step give instant access to the operator's platform and with a convenient hand grip on the wheel fender; you can swing yourself up to the seat in a trice without stumbling over close coupled or three-point link and machines.

**Other Major Features include:** a double spool valve as regular equipment allowing independent or simultaneous operation of three-point linkage and double acting remote control system, sealed beam headlights mounted on rubber in an attractively designed grille; magnetic traps installed in both transmission oil galleries, which afford 99% bearing protection from metal materials so greatly increasing bearing life; independent PTO which can be engaged on the move; wide swinging drawbar allowing easy turns when working with trailed implements.

**Independent Hydraulics** The engine-driven pump provides instant hydraulic power inde-

pendently of clutch operation and smooth working of three-point linkage or remote control is ensured with the accessible, easy-to-operate double spool valve. This is attached to the steering column by your right hand. The valve enables both linkage and remote control to be actuated independently or together.

**564 Farmall Tractor**

Introduced in 1969, the 564 Farmall with wheel tracks adjustable to 100 inches is the ideal tractor for all row-cropping. With an improved hydraulic system and three-point linkage, the 564 International Farmall has precision control over a full range of mounted equipment. New lower linkage arm mounting pads and inboard lateral controls ensure even greater crop clearance than previously. Double spool valve allows independent operation of mid-mounted or trailed equipment separately to 3-point linkage. This is without doubt the most versatile row crop tractor available.

**High Clearance**

Exceptional rear axle clearance of 26 inches enables you to cultivate to a late stage of crop growth without damaging plants. The 564 has a turning radius of 99 inches with brakes applied and 102 inches without (less than its own length) to allow more work to be done through shorter headlands and time to be saved.

**Fits all Crops**

The adjustable rear wheels permit the 564 to fit any crop. Standard rear wheel axles give a range of adjustment from 52 inches to 100 inches. Rear wheel track is simply altered by sliding wheels along their keyed axles. For the front, a single wheel or an adjustable wide axle to give a track from 57 to 87 inches is available instead of the standard twin wheels.

**564 Diesel Tractor Specifications**

**Horsepower**

| | |
|---|---|
| Stripped HP | 61 |
| Brake | 55 |
| Max belt. | 51.7 |
| Max Drawbar | 50.8 |

**Engine**

| | |
|---|---|
| Number of cylinders | 4 |
| Bore and Stroke (inches) | 4 x 5.25 |
| Piston displacement (cu. in) | 264 |
| Speed at rated engine rpm | 1800 |
| Clutch Size (inches) | 12 |
| Cooling System type | Pump and pressure |
| Lubrication type | Full Pressure |

**Belt Pulley**

| | |
|---|---|
| Diameter and face (inches) | 11 x 7 ½ |
| Belt speed (fpm) | 3,035 |

| **Speeds MPH** | **16.9 x 30 rear tyres** | **18.4 x 30 rear tyres** |
|---|---|---|
| 1st Gear | 2.4 | 2.6 |
| 2nd gear | 3.5 | 3.7 |
| 3rd gear | 4.5 | 4.8 |
| 4th gear | 5.4 | 5.8 |
| 5th gear | 15.2 | 16.0 |
| Reverse | 3.2 | 3.4 |

**Independent PTO**

| | |
|---|---|
| Speed at rated engine RPM | 543 |
| Spline shaft Diameter (inches) | 1⅜ |

**Tyre Sizes**

| | |
|---|---|
| Front | 7.50 x 18 |
| Rear (standard Axle) | 16.9 x 30 |

**Wheel Treads (in)**

| | |
|---|---|
| Front | 52⅜ |
| Rear – 16.9 x 30 tyres | 61 |
| - 18.4 x 30 tyres | 63 |
| Wheelbase | 76 ½ |

**Turning radius (in)**

| | |
|---|---|
| With brakes applied | 132 |
| Without brakes applied | 156 |

**Ground Clearance**

| | |
|---|---|
| Under front axle | 14 ¼ |

| | |
|---|---|
| Length – overall (in) | 125 |
| Width – on 16.9 x 30 tyres | 78 |
| - on 18.4 x 30 tyres | 80 ½ |
| Height to top of radiator (in) | 60 ½ |
| Weight -with fuel, operator and water ballast | 8,160 pounds |

**Capacities** (approximate Imperial measures)

| | |
|---|---|
| Fuel tank | 17 ½ gallons |
| Water cooling system | 5 gallons |
| Crankcase Pan | 13 pints |
| Transmission case | 10 ¾ gallons |
| Air cleaner-oil cup | 2 pints |
| Steering gear housing | 1 ¾ pints |

**Standard Equipment**

Diesel engine; 12 volt electrical system; tractor meter; voltage regulator; heavy-duty co-axial starter motor; oil bath air cleaner; floating oil screen; replaceable-element oil filter; muffler-spark arrester; enclosed external contracting foot brakes operation on differential shaft; spring loaded foot clutch; engine driven hydraulic system with double spool valve; vertically adjustable rear roller drawbar; super deluxe seat with arm rests/ back rest; tool box; pre-screener; pre-cleaner; independent power take-off; 7.50-18, 6-ply front tyres; 16.9-30, 6-ply rear tyres on DW-14 rims; front wheel weights (set of 2); 1st rear wheel weight (set of 2); lighting (2 front and 1 rear lamp); power steering.

**Optional Equipment**

Belt pulley; category 2 3-point linkage; hydraulic remote control unit; second/third rear wheel weights (set of 2); 16.9-30 8-ply tyres on DE-14 rims; 18.4-30 6-ply tyres on DW-16 rims; Foot throttle.

*Prototype air cleaner for 564*

**Special Features**

A magnetic plug installed into the transmission oil galleries prevents foreign particles from damaging the transmission bearings. A new radiator grille with a steel grille insert and mesh readily removable, constitute easy cleaning. Power steering and a Super Deluxe padded seat with back rest and arm rests are regular equipment for the 564 Farmall.

**Specifications; Farmall 564**

| **Horsepower** | 61 |
|---|---|
| Fully equipped HP | 55 |
| Max. Belt | 51.7 |

**Engine**

| Number of cylinders | 4 |
|---|---|
| Bore and Stroke (in) | 4 x 5.25 |
| Piston displacement (cu. in) | 264 |
| Speed at rated engine rpm | 1800 |
| Clutch Size (inches) | 12 |

Cooling System type: Pump and pressure

Lubrication type: Full Pressure

**Belt Pulley**

| Diameter and face (inches) | 11 x 7 ½ |
|---|---|
| Belt speed (fpm) | 3,030 |

| **Speeds MPH** | **13.6 x 38 rear tyres** | **14.9 x 38 rear tyres** |
|---|---|---|
| 1st Gear | 2.6 | 2.7 |
| 2nd gear | 3.8 | 3.9 |
| 3rd gear | 4.9 | 5.0 |
| 4th gear | 5.9 | 6.1 |
| 5th gear | 16.6 | 17.1 |
| Reverse | 3.5 | 3.6 |

**Independent PTO**

| Speed at rated engine RPM | 543 |
|---|---|
| Spline shaft Diameter (inches) | 1 ⅜ |

**Tyre Sizes**

| Front | 6.00 x 16 |
|---|---|
| Rear (standard Axle) | 14.9 x 38 |

**Wheel Treads (in)**

| Front | 8 ½ |
|---|---|
| Rear (standard axle) | 52 to 100 |

**Turning radius (in)**
With brakes applied 99
Without brakes applied 102
**Ground Clearance**
Under rear axle 27 ¾
Under rear frame 19 ¼
Length – overall (in) 139
Width – (outer ends of 100in axles, in) 98
Height to top of steering wheel (in) 82 ¾
**Swinging drawbar adjustment**
Lateral, each side of centre (in) 8 ⅜
from ground 11⅞, 14, 16 ⅛, 18 ¾

**Capacities**
(approximate Imperial measures)

| | |
|---|---|
| Fuel tank | 17 ½ gallons |
| Water cooling system | 5 gallons |
| Crankcase Pan | 13 pints |
| Transmission case | 10 ¾ gallons |
| Air cleaner-oil cup | 2 pints |
| Steering gear housing | 1 ¾ pints |

**Standard Equipment**
Diesel engine; 12 volt electrical system; tractor meter; voltage regulator; heavy duty co-axial starter motor; oil bath air cleaner; floating oil suction screen; replaceable element oil filter; five forward speeds, one reverse; muffler-spark arrestor; heavy duty external contracting band type independent foot brakes; spring loaded clutch; engine driven hydraulic system with double spool valve; power steering; roller swinging drawbar; toolbox; air-cleaner pre-screener and pre-cleaner; three-point linkage; 6.00 -16 front tyres, 14.9-38 wide base 6-ply rear tyres on demountable steel rims; front and rear wheel weights (1 set each)

**Optional Equipment**
Belt pulley; independent power take-off; electric lighting- two front and one rear; additional wheel weights; two-way remote hydraulic control; 106 inch tread rear wheel attachment; deduct for 13.6 x 38 tyres; front end weights.
For specialized equipment an adjustable wide front axle wheel can be fitted.

*Above: Studio photo of Standard 564*
*Below: Photo of Farmall 564.*

**564 B Series Tractor**

In 1970 the 564 was upgraded to the series B. Many improvements and refinements made it necessary to introduce this proven and improved tractor. Early B Series used decals to label the "B Series" under the 564 emblem. Later models have the "B Series" integrated into the 564 emblem. Farmall 564 B Series tractors were also sold, although few were built. The specifications for the Farmall 564 B are the same as the Farmall 564.

**Forged from Experience**

The 564 Series B, with its 264 cubic inch displacement engine is not a new concept with unknown faults, hidden weaknesses, last minute improvisations. The 564 Series B is a tractor that has been improved until perfected; strengthened in places where increased strength was needed to cope with today's intensive, high-speed farming, re-designed for greater comfort, greater operating ease and greater efficiency. Improvements are added, not to improve the face value, but to evolve and perfect the International 564 Series B.

**Scientific weight distribution**

Under normal conditions, wheel spin becomes a thing of the past. Scientifically designed to put weight where it is needed, the 564 Series B with its wide tyres, provides the flotation necessary to ride over soft, wet ground; to haul implements that would stall other tractors in its class.

**Top in Comfort and Economy**

The 564 Series B offers unbeatable traction produced by a four-cylinder 61hp engine, unbeatable comfort to make those long working hours pass quicker in a deluxe, foam padded seat which absorbs all shocks and bumps, unbeatable fuel economy using a modern distributor-type pump that saves on those heavy fuel bills - In short, an unbeatable tractor.

**Specifications; 564B**

| | |
|---|---|
| **Horsepower** | |
| Stripped HP | 61 |
| Brake | 55 |
| Max. Belt | 51.7 |
| Max. Drawbar | 50.8 |
| **Engine** | |
| Number of cylinders | 4 |
| Bore and Stroke (in) | 4x5.25 |
| Piston displacement (cu. in.) | 264 |
| Speed at rated engine r.p.m. | 1800 |
| Clutch Size (in) | 12 |
| Cooling System (type) | Pump and pressure |
| Lubrication (type) | Full pressure |
| **Belt Pulley** | |
| Diam. and face | 11x7 ½ |
| Belt speed (f.p.m.) | 3035 |

| Speed MPH | 16.9x30 Rear Tyres | 18.4x30 Rear Tyres |
|---|---|---|
| 1st gear | 2.4 | 2.6 |
| 2nd gear | 3.5 | 3.7 |
| 3rd gear | 4.5 | 4.8 |
| 4th gear | 5.4 | 5.8 |
| 5th gear | 15.2 | 16.0 |
| Reverse | 3.2 | 3.4 |

**Independent PTO**

| | |
|---|---|
| Speed at Rated engine rpm | 543 |
| Spline shaft diam (in) | 1 ⅜ |

**Wheel treads (in)**

| | |
|---|---|
| Front | 52 ⅝ |
| Rear – 16.9 x 30 tyres | 61 |
| 18.4 x 30 tyres | 63 |
| Wheelbase (in) | 76 ½ |

**Turning Radius (in)**

| | |
|---|---|
| With brakes applied | 132 |
| Without brakes applied | 156 |

**Minimum Ground Clearance (in)**

| | |
|---|---|
| Under front axle | 14 ½ |
| **Length** – Overall | 125 |
| **Width** - overall (in) | |
| On 16.9-30 tyres | 78 |
| On 18.4 -30 tyres | 80 ½ |
| Height to top of radiator | 60 ½ |
| Weight with fuel, operator and water ballast (lb) | 8,160 |

**Capacities (approx. Imperial measures)**

| | |
|---|---|
| Fuel Tank | 17 ½ gal |
| Water cooling system | 5 gal |
| Crankcase pan | 13 pints |
| Transmission Case | 10 ¾ gal |
| Steering gear housing | 1 ¼ pints |

**Tyre Sizes**

| | |
|---|---|
| Front | 7.50 x 18 |
| Rear | 16.9 x 30 |

*Above: 564B on Assembly line*

*Close-up View of 564B Air Cleaner.*

*Product Engineering photo of 564B —notice battery placement.*

**Standard Equipment**

Diesel engine; 12 volt electrical system; tractor meter; voltage regulator; heavy-duty co-axial starter motor; oil bath air cleaner; floating oil screen; replaceable-element oil filter; muffler-spark arrester; enclosed external contracting foot brakes operation on differential shaft; spring loaded foot clutch; engine driven hydraulic system with double spool valve; vertically adjustable rear roller drawbar; super deluxe seat with arm rests/ back rest; tool box; pre-screener; pre-cleaner; independent power take-off; 7.50-18, 6-ply front tyres; 16.9-30, 6-ply rear tyres on DW-14 rims; front wheel weights (set of 2); 1st rear wheel weight (set of 2); lighting (2 front and 1 rear lamp); power steering.

**Optional Equipment**

Belt pulley; category 2 3-point linkage; hydraulic remote control unit; second/third rear wheel weights (set of 2); 16.9-30; 18.4-30 tyres; Foot throttle.

564 was advertised from 1966 to 1969
Farmall 564 advertised from 1968 to 1973
564B advertised from 1970 to 1973

*Below: Pre-production photo of 564B*
*Photo dated 16 April 1969*

**Tractor Build Data**
**Aug 1966 – May 1972**

2,364 Units Built – Production numbers are not available for the individual tractor variations (Standard or Rowcrop) Listing also does not designate when tractor was switched over to 'B' Series. In 1970, according to Power Farming Australia books, the B series was introduced.

| | |
|---|---|
| 564 | 1969 price $3,595 |
| Farmall 564 | 1972 price $3,575 |
| 564 B Series | 1972 price $3,675 |

Australian Tractor test number 46 – May, 1967

| Financial Year (July 1 – June 30) | Units |
|---|---|
| 1966/1967 | 802 |
| 1967/1968 | 672 |
| 1968/1969 | 444 |
| 1969/1970 | 204 |
| 1970/1971 | 149 |
| 1971/1972 | 93 |

For monthly production list, see Appendix D.

# Chapter 10
# 434

## 434

The McCormick International 434 is another great tractor, replacing the A-414 which won universal acclaim and praise throughout Australia. Australian built 434 is the first tractor built in Australia with a Speed Amplifier, commonly known as a Torque Amplifier or TA for short. The 40 horsepower Australian built 434 was sold form November 1966 to August 1972. The 434 Hi-Clear was introduced in 1969.

In superseding the A-414 the 434 is offering more, much more. More of what? More of everything you'd want in a 40 hp tractor. It's built to do more work each day, more versatile jobs and give more years in its working life.

How can the 434 offer you so much more? To begin with, it's built on experience, nobody in the world has as much experience of tractors as International Harvester. Equally important, it's designed with more thought and engineered with more care – like all IH tractors.

Big hearted power means power at all speeds and that's what you get from the well proven International AD 154 40hp diesel engine. It provides power when you need it most – for tackling big jobs under tough conditions, for shifting big loads in double quick time.

Power for the hydraulics, power for the PTO – the AD154 supplies it reliably, dependably, day in, day out. The unique IH differential lock with its robust and positive action operates on the simplest principle. Even if a wheel has started to spin, it operates faultlessly every time without reducing power. A slight pressure from the right foot and the diff lock give you grip on both wheels. No grating or grinding.

The handbrake with over centre lock has positive action, either on or off. There is no halfway position.

The lever has a convenient fingertip adjustment. Sealed beam headlights are mounted in the radiator grille for maximum protection and securely mounted on a solid bolster for accurate positioning. They are simple to adjust and are protected from dirt and corrosion.

The grille comes off in a flash to expose the air cleaner for oil level checks and filter cleaning. In this position the filter is exposed to clean cool air and gives higher filter efficiency. Constant running 545 rpm power-take-off is regular equipment. Gear changes can be made or the tractor stopped, without affecting the PTO operation.

*McCormick International 434 tractor with A1-41 disc plow is put through paces as part of practical use of new tractors and three point linkage implements*

Above: *434 at work*
*Below: Demonstrating how easy it is to check water, fuel and oil*

**More maintenance accessibility**
Lift one cover and you can top up oil, fuel and water. A good view of water and fuel levels means you can check at a glance – and reduces risk of spillage.

**IH Speed Amplifier** provides a speed for every conceivable job. It also enables you to obtain a lower speed and increased pull, on the go, without stopping or declutching. Hydraulic power does it for you – smoothly and without hesitation.

**IH Vary-Touch hydraulics** enables the tractor to think for itself. They incorporate the latest methods of operation linkage mounted implements or single and double-acting remote control cylinders.

**Specifications**
**Engine:**

Four-stroke

**Tractor Speeds**
(at rated governed speed on 13.6 x 28 tyres)

| | Standard | Speed Amplifier |
|---|---|---|
| L1 | 1.46 | 2.01 |
| L2 | 2.50 | 3.44 |
| L3 | 3.56 | 4.89 |
| L4 | 5.31 | 7.30 |
| H1 | 4.26 | 5.85 |
| H2 | 7.28 | 10.00 |
| H3 | 10.37 | 14.25 |
| H4 | 15.49 | 21.25 |
| Low reverse | 2.22 | 3.05 |
| High reverse | 6.46 | 8.88 |

diesel, valve –in-head with electric all-weather glow-plug starting.

IH Model AD-154

Horsepower (stripped) 40
Number of cylinders 4
Bore and Stroke 3 ½ x 4
Piston Displacement (cu. inches) 154
Rated governed speed (rpm) 2,000
Maximum torque (lbs. ft) 115
RPM at maximum torque 1,300

**Electric starting and Lighting:** 12 volt electric starting system which includes key-start, solenoid-engaged started motor, glow plugs, resistance indicator, generator, voltage regulator, one 12-volt battery, 12-volt front and rear lights.

**Engine Clutch:**
Dual clutch – 2-plate clutch, heavy-duty friction linings and loaded by heavy duty Belville washers
Transmission clutch diameter: 11 inches
PTO clutch diameter: 9 inches

**Standard Transmission:**
Eight forward speeds and two reverse selected by a gear lever and a high-low shift lever.

**Optional Transmission:**
As listed but with Speed Amplifier giving 16 forward, four reverse speeds.

**Brakes:**
Foot operated, totally enclosed, double disc, externally mounted, acting on bull pinion shafts, one on each side. Foot pedals latch together for road work and a hand parking brake is provided.
Disc diameter 6 ½ inches
Effective friction area (each lining) 23.56 sq. inches
Number of linings; each side 4
Total effective friction area 188.48 sq inches

**Hydraulic System:**
Independent fully 'live' system with Vary-Touch draft and position control. Average lifting force at end of lower links 2,600 pounds

**Tractor Dimensions (inches)**

**Length, overall** 118 ¼

Tread, front (in 4 inch stages) 52 to 76

Tread, rear 53 to 76 ½

**Width, overall:**

Front Wheels 59 ½ to 87 ½

Rear Wheels 64 ½ to 88

Height to top of steering wheel 59 inches

Wheelbase 76 ½

Ground Clearance (min. under frame) 17 ¼

Ground Clearance (min. under front axle) 18 ¾

Turning Radius:

At minimum setting 133 inches

with brakes applied 114 inches

**Capacities**

| | |
|---|---|
| Cooling system | 2 ½ gallons |
| Fuel tank | 10.6 gallons |
| Engine lubrication | 13 pints |
| Transmission case | 4 ¾ gallons |
| Hydraulic reservoir | 2 ½ gallons |
| Air cleaner cup | 1 ¼ pints |

**Operating Weight** Basic tractor, including fuel, oil, water, hydraulic system with three-point linkage and constant running PTO 4,900 pounds

**Standard Equipment** 3-point linkage; foot operated differential lock; sealed beam headlights and tail light; heat indicator; safety key start; tractor-meter; deluxe suspension seat; constant running 545 rpm PTO; eight-speed transmission, tyre equipment; 6.00 x 16 front, 13.6 x 28 rear

**Optional Equipment** Speed amplifier giving 16 forward speeds and four reverse; two speed (545 and 745 rpm) constant running PTO (not available with Speed Amplifier); single clutch 545 rpm PTO; belt pulley; down swept ex-

Showing the park brake capabilities

haust; self-sealing coupling; front and rear wheel weights; single or double-acting remote control attachment; swinging drawbar and frame; 3-point linkage drawbar; tyre equipment: 7.50 x 16 front, 13.6 x 28 rear, 14.9 x 28 rear, 14.9 x 24 rear, 16.9 x 28 rear.

**Tractor Build Data**

Nov 1966 – Aug 1972

| Cost in 1970 | $2,888 |
|---|---|
| Hi-clear | $2,970 |

Australian Tractor Test Number 47, June 1967

5,446 Units Built—Production numbers are not available for the individual tractor variations. (Standard or Hi-Clear)

**Financial Year (July 1-June 30)**

| **Year** | **Units** |
|---|---|
| 1966/1967 | 776 |
| 1967/1968 | 1251 |
| 1968/1969 | 816 |
| 1969/1970 | 983 |
| 1970/1971 | 780 |
| 1971/1972 | 740 |
| 1972/1973 | 100 |

For monthly production list, see Appendix D.

**434 Half-Track Attachment**

Where increased traction, increased stability and extra floatation is desired, the International half track attachment fitted to a 434 tractor is just the tool. Two types of belts are available; all steel and rubber linked belts.

All steel tracks are built and recommended for extra heavy duty service in average soil conditions. Plenty of traction between tyre and tracks is assured because there is no stretch, even under full load. In case of accidental damage, replacement links can be quickly installed at a low cost.

Rubber belted tracks are most satisfactory where conditions are very abrasive. Belting is of highest tensile strength and is specially made to resist stretching. Shoes are interchangeable with steel linked tracks and extend over the edges of the rubber belting to prevent scuffing against rocks, tree stumps, and other hard cutting surfaces that keeps maintenance costs low.

Quickly detachable idler wheels keep tension on the belts.

***Forestry crews were given a chance to test their own equipment with IH machines. Here a half-track conversion of McCormick International 434 tractor works with a tree planting unit.***

**434 Hi-Clear**

The 434 Hi-Clear has a 26⅝ inch clearance under the front axle, allowing the tractor to cultivate row-crops to a late stage of growth. Front wheel treads from 55 to 76 ½ inches, enabling the tractor to suit all types of crop. A small turning radius of 114 inches, using brakes, gives accurate control for tight field work.

Designed especially as a high clearance row-crop tractor, the 434 Hi-Clear also incorporates all the popular design features of the regular 434. Three point linkage with live hydraulics with "Vary Touch" control for position and draft control leads in precise easy operation of a variety of mounted implements.

Standard wheel equipment: 6.00 x 16 front tyres; 12.4 x 38 rear tyres.
Optional rear tyres, 13.6 x 38

**434 Extra High Clear**

This is a prototype model and was not manufactured. The front and rear axle hub on this tractor was dropped four inches.

Above and below are three photos dated 28 February 1967 showing the extra length in the front axle, the altered rear axles and the larger tyres used for this special tractor with the normal 16" front wheels next to it for comparison.

**434 Lowboy**

Found in the Product Engineering files, this tractor complements the Extra High Clear 434—by going as low as possible.

Dated 2 September 1966, these photos are a great example of the engineers listening to the farmers needs—in this case fruit and orchard growers. With low profile and down swept exhaust, this tractor has it all. Too bad it was not available to the general public.

**434 Special**

Another Engineering project that never came to light, this is dated 4 July 1967, this tractor seems to be like a regular 434, but with wide fenders.

# Chapter 11
# 686/756

The Australian-produced 756 and the Australian 686 are necessities for high-speed, broad acre farming. These tractors will keep churning out power hour after hour, day after day. Solidity is a key word of IH tractors and these giants, with their massive axles and axle carriers and exclusive Elotherm induction hardened crankshaft are no exception.

The 686 and 756 tractors introduce new standards to the farming world. An adjustable wide front axle is standard equipment, with three adjustments, 56, 69 or 80 inches—a size to suit every ground condition.

Hydrostatic power steering, the first Australian built tractor with this feature, is the easiest, surest steering there is—with no mechanical linkage. Steering is actuated by a double acting hydraulic cylinder in lieu of a conventional drag link. The steering wheel meters the oil flow to the hydraulic cylinder. Even with the engine off, the steering wheel actuates the oil which gives the conventional steering effect. The wheel shocks are absorbed by the oil cushion giving shock free steering to the operator.

The International 756 has been designed for farmers who don't want to hold back when they are pulling heavy disc ploughs and big broad acre cultivators. This is the tractor that's built to pull at speeds that spell the difference between getting the work done fast and just getting it done.

Power to weight ratio is excellent for top performance on every job. And you get 16 speeds with exceptionally easy gear shifting, plus shift on the go torque amplifier, as standard.

Differences of the 686 and 756 is minimum, the main difference is the 686 is a 89 horsepower tractor with eight standard speeds and the 756 is a 101 horsepower tractor with standard 16 speed, shift-on-the-go torque amplifier.

The 6-cylinder 310 cubic inch International diesel engine for both tractors is big, tough and rugged and the low friction design has proved itself in many hours of high-speed, hard-pull farming. Dyna-life clutch, armchair seating, and remote controls to breakaways are some of the features of the 686 and 756 tractors. Three-point linkage equipment is available for both tractors as an option and one of three separate hydraulic pumps enables a constant source of hydraulic power through the IH Torsion bar lower link sensing Draft control.

## 686

**Engine:** International D-310 direct injection diesel; 6 cylinders; 3.88 inch bore; 4.38 inch stroke. Displacement 310 cubic inches
Engine Max Horsepower—89
PTO horsepower -64.0
Drawbar horsepower – 61.0

**Transmission:** 8 forward, 4 reverse with high/ low range

**Hydraulics:** reservoir capacity—18.75 gallons, pump capacity—10 gpm, max pressure 2,100psi

**Standard speeds:** (mph) on 18.4 x 34 rear tyres

Forward: 1.7, 2.3, 4.1, 5.4, 6.2, 8.1, 14.2, 19.0
Reverse: 2.8, 3.7, 6.4, 8.6

**Standard tyres:** Front—7.50 x 20;
Rear—18.4 x 34
**Optional tyres:** Front—10.0x16
Rear—18.4x30 duals or 23.1x30
**Weight**:
Dry: 9,268 pounds

## 756

**Engine:** International D-310 direct injection diesel; 6 cylinders; 3.88 inch bore; 4.38 inch stroke. Displacement 310 cubic inches
Engine max horsepower—101
PTO horsepower – 76.0
Drawbar horsepower – 69.0

**Transmission:** 16 forward, 8 reverse with Torque Amplifier and hi/low range

**Hydraulics:** reservoir capacity—18.75 gallons, pump capacity—10 gpm, max pressure 2,100psi

**Standard speeds:** (mph) on 18.4 x 34 rear tyres
Forward: 1.5, 2.25, 3.75, 5, 5.75, 7.75, 13.25, 17.75
With TA: 1.25, 1.75, 3, 4, 4.5, 6, 10.5, 13.75

**Standard tyres:** Front—10.0x16
Rear—23.1x30

**Optional tyres:** Rear—18.4x30 duals
18.4x34

**Weight:** Dry: 9,390 pounds

*Australian design and manufacture of farm equipment has a long and proud history. Australians in the past have made positive contributions to world agriculture with inventions such as the stripper, stump jump plough and spring-tyne cultivator. Now Australian expertise in equipment design and manufacture is illustrated by these machines shown—the International 1-6 disc plough drawn by an International 686 tractor.*

**Specifications for both 686 and 756**

**Dimensions:**

Turning radius: 10.9 feet without brakes
Ground clearance 13.0 inches.
Wheelbase: 105inches
Length: 164 inches
Width: 92inches
Height to top of steering wheel: 82 inches
Tread, front: 57 to 81 inches
Tread, rear: 60 to 104 inches

**Capacities**

Cooling system: 23.5 quarts
Fuel Tank: 27.5 gallons
Engine lube: 12 quarts
Transmission and Diff: 17 gallons

**Instruments and warning lights**

lights—two front grille mounted with combination rear stop/tail light.

**Implement attachments**: optional; Cat 2 linkage, max lift at hitch point: 10,000 pounds, clutch housing mounting pads.

Cabin or Rollbar: weather cab available

Additional equipment available: front end weights, rear wheel weights, double acting remote control valve, foot pedal accelerator, front end mounting bracket and weights.

**Tractor Build Data**

The Australian built International 686 and 756 were built under the same serial numbers. There is no distinction in the build sheets how many of each were built.

Oct 1969 – June 1971
830 Units Built - Production numbers are not available for the individual tractor variations.

| | |
|---|---|
| 1968/1969 | 467 |
| 1969/1971 | 359 |

For monthly production list, see Appendix D.

| | |
|---|---|
| 686 Price in 1970 | $5,990 |
| 756 Price in 1970 | $6,990 |

Australian Tractor Test Number 80—for 696 tractor. Nebraska Tractor Test 955 for 756 tractor was used for comparison. November, 1966.

# Chapter 12
# 696/766

The 696/766 tractors replace the 686 and 756 offering more strength and weight than the previous models. Designed for broadacre use to pull disc ploughs, scarifiers, cultivators and seed drills the D-310 diesel engine gives all the pull to deliver a full day's work.

The two tractors build sheet data are the same as well as the serial numbers. The difference between the 696 and 766 is the torque amplifier – only available on the 766. The 766 also features the 'black stripe' decals in the late build model. Offering 89 horsepower for the 696 and 98 horsepower for the 766, these tractors have a super deluxe seat, finger-tip power steering, power disc brakes and a full range of instruments. The wide swinging roller drawbar with a vertically adjustable tongue, double acting remote control with double acting check valve and breakaway coupling is standard.

Dual speed independent PTO with rear working light, tail and stop light as well as the 766 also offers a rear operators deck for more room and easy access.

Linkage is optional equipment on a tractor this size and the IH three point linkage, with up to 6800 pounds of lift, gives you torsion bar lower link sensing to provide a fast weight transfer allowing you to get out of trouble before you get into it. Automatic traction control allows for as small as a 5% load change to put more or less weight on the rear wheels as needed.

Easy on easy off handgrips on the fender and engine cowling and two well placed steps make forward mounting a 696 or 766 easy. All levers are off the deck and out of the way and the seat can be moved out of the way for standup driving.

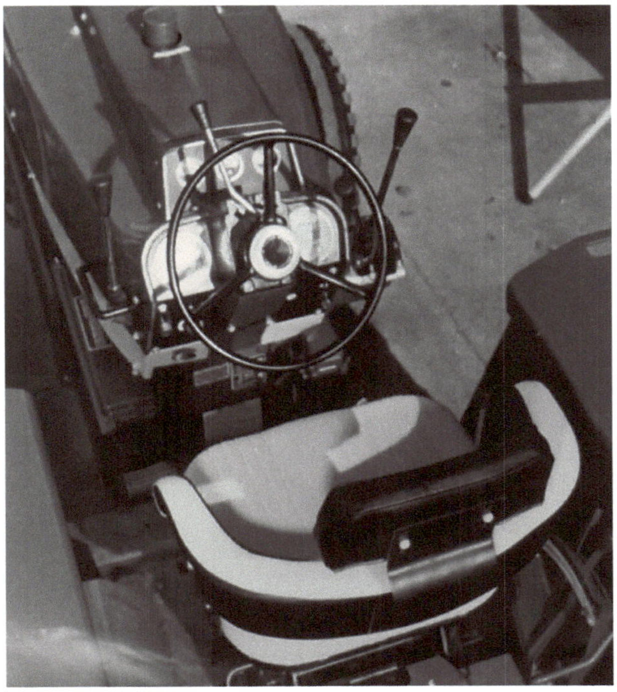

Wide, spacious and free from obstructions it's the easiest deck to get on and off.

Gauges, not lights, keep the operator informed. The upper bank of instruments includes air cleaner service indicator, fuel gauge and ammeter. At lower right are engine oil pressure gauge and engine water temperature gauge. On the left is the speedometer- tachometer. Transmission oil pressure is indicated by a warning light. Below these there is a cigarette lighter, light switch key and key start. Gear shift on the right, Torque Amplifier lever on the left (only on 766) are off the deck and easy to use. All dials and levers are illuminated by indirect light and a safety feature is the engine cannot be started unless the clutch pedal is depressed.

Hydrostatic power steering gives the driver comfort and safe driving. With an external hydraulic cylinder directly connected to the centre steering arm, there are fewer parts to service and steering takes less effort. It's smooth, all hydraulic and allows for steering even with the engine off.

Isomount Isolators are a new result of IH research in tractor comfort. The isolators are specially designed rubber mountings that are highly effective in absorbing vibration. They are used in mounting the operators' platform, fenders, steering column, gear selector support, fuel tank and sheet metal to the basic tractor. These isolators completely eliminate metal to metal contact and by isolating various parts that tend to conduct or amplify vibration, these Isomount isolators substantially reduce vibration entering the operators' area.

Swing out oil cooler permits quick inspection and cleaning while the air cleaner pre-cleaner baffle ahead of the radiator deflects most of the dust and dirt before it can reach the air filter. Engine fan pulls dust-laden air through the radiator, baffle directs clean air up and over the engine, which greatly reduces the frequency of air filter cleaning.

Double- disc dry external power disc brakes are smooth and sure for stops and turning assistance. Brakes automatically equalize and self adjust for wear with curved pedal faces for comfortable operation.

Dry-type dual-stage air cleaner assures 99.98% efficiency. Filter has an automatic dust unloader plus cleanable element with a final built-in safety element to keep dirt out. No oil, no mess and works with pre-cleaner baffle located ahead of radiator. Electric service indicator tells when element needs cleaning.

The IH D-310 engine features multi-range power, hydraulically governed fuel pump, direct injection, water jet cooled valves, cross flow cooling, full pressure lubrication that add up to a potent power package designed to give maximum tractor performance. Dual spin-on, full flow, throw-away filters remove the finest impurities to protect the pump and injector nozzles.

Lever linkage on the 696 and 766 allows for easy changing of gears, the "H" pattern provides neutral between HI and LO ranges. Eight forward speeds include two creepers, four speeds for fast tillage and two for fast transport. 766 also comes equipped with the Torque Amplifier. This doubles the number of speeds to 16 forward and 8 reverse and lets you power shift on the go.

**Specifications**

**Engine**

International D-310, 6 cylinder, Direct Injection

| Piston displacement | 310 cubic inches | |
|---|---|---|
| | **696** | **766** |
| Horsepower | Max 90 | Max 98 |
| Rated governed speed | 2300 | 2300 |
| Max torque at 1600rpm | 227 lbs/ft | 248 lbs/ft |
| Air Cleaner | Dry | Dry |
| Drawbar horsepower | 61 | 69 |
| **Engine Clutch** | 4 Button; "Dyna-Life" | |
| Diameter | 12 inches | |
| **Transmission** | Sliding Gear | Sliding Gear Torque Amplifier |
| Speeds | 8 forward, 4 reverse | 16 forward, 8 reverse |
| Final Drive | Spur gear | Spur Gear |
| PTO Independent | Dual | Dual |
| Speeds | 540 and 1,000 | |
| Engine rpm | 2,100 | 2,100 |
| PTO HP | 70 | 78 |

**Steering**

| Type | Hydrostatic Power |
|---|---|

**Brakes**

| Type | Power Double Disc |
|---|---|
| Size | 8" o.d. x 5" i.d. |

**Electrical**

| Batteries | 2, 6-volt |
|---|---|
| Safety Starter | Key switch, clutch safety switch, push button start |

**Hydraulic system**

| Reservoir capacity | 18.75 gallons |
|---|---|
| Pump type | Gear |
| Pump rated output | 10gpm |
| Pressure (Max.) | 1,550 lb. sq. in. |
| Linkage (optional) | Cat 2 |
| Lift at link ends | 6,800lbs |

**Capacities**

| Fuel | 35 gal |
|---|---|
| Cooling system | 6.2 gal |
| Engine Oil | 2.75 gal |
| Transmission Case | 18.75 gal |

| **Weight** | **696** | **766** |
|---|---|---|
| Shipping weight | 9,630lb | 9,860lb |
| Maximum weight | 13,600lb | 14,300lb |

**Dimensions**

| Length | 155.2in | 155.8in |
|---|---|---|
| Wheelbase | 104.8in | 104.8in |
| Width | 78.4in | 91.0in |

## Instrumentation

**Gauges** – air cleaner indicator; fuel ammeter; oil pressure; temperature; speedometer-tachometer; transmission oil pressure warning light.

**Lighting**

Two headlights Grille mounted; rear flood light; combination stop and tail light; indirect light for instruments; Optional front lights mounted in Rear Fenders

**Tractor Build Data**

Mar 1972 – June 1978

4,271 Units Built

| | |
|---|---|
| 1971/1972 | 368 |
| 1972/1973 | 518 |
| 1973/1974 | 692 |
| 1974/1975 | 795 |
| 1975/1976 | 780 |
| 1976/1977 | 727 |
| 1977/1978 | 369 |

For monthly production list, see Appendix D.

| | | |
|---|---|---|
| Height to steering wheel | 84.2in | 84.7in |
| **Ground clearance** | | |
| Under drawbar pivot | 14.2in | 14.9in |
| Under front axle | 23.6in | 22.6in |
| Turning radius with brakes | 132in | 132in |
| Without brakes | 163in | 163in |

Front axle adjustment **(696)** 57-81" in 4" increments **(766)** 60-84" in 4" increments

Rear axle adjustment 60-99in 68-103in

Roller drawbar swing w/out hitch, both models, 40in

**Tyres**

| | | |
|---|---|---|
| Standard – rear | 18.4 x 34 | 23.1 x 30 |
| Standard – front | 7.5 x 20 | 10.0 x 16 |
| Optional- Rear | 18.4 x 34 | 18.4 x 34 |
| | 23.1 x 30 | 23.1 x 30 |
| | Dual | Dual |
| | 18.4 x 30 | 18.4 x 30 |
| Dual | Dual | |
| Optional – Front | 10.0 x 16 | |

# Chapter 13
# 444

The International 444 replaces the International 434 tractor. This tractor has the same engine, weight, transmission and general configuration as the A-414 and 434 model tractors.

A tractor built for general runabout jobs, pulls balers, mowers, mounted ploughs, cultivates between the vines, trees, vegetables, and does just about anything you need.

The work orientated styling will catch your eye first but take a closer look and you will notice a host of features on this utility tractor. Comprehensive instrumentation, big, easy to handle steering wheel, fuel tank tucked out of the way and not over the engine, foot throttle, dry type air cleaner, safety starter switch, protection and ease of separate water trap, front axle support casting.

Powered by AD-154 diesel engine, eight forward, two reverse speed transmission, six matched work speeds and two road speeds, well placed hydraulic controls, generously cushioned suspension seat with tension springs to suit different operator needs, and hydraulic shock absorber to eliminate bumps and jars.

It is easier to farm with the 40 horsepower 444 tractor because it was designed that way. The control and feel of the foot throttle, easy all-weather glow plug starting and dependable power from a proven engine in tractor applications. Ground speeds designed to handle the range of jobs for the 444 with eight forward and two reverse with a differential lock to walk out of those occasional tough spots, eliminating wheel spin. Totally enclosed, externally mounted 6 ½ inch diameter self energizing double disc brakes gives sure safety of stopping.

Fully live hydraulics with big capacity pump and vary-touch system gives full power when the engine is started, independent of the transmission or engine clutch. Sensitive operation absorbs and transfers implement mass and draft to the tractors rear wheels for better traction.

Multi-adjustment linkage gives excellent control over your field work, the 444 system incorporates seven individual adjustment points, quick-hitch attachments, different soil conditions and different implements are all provided for.

Full accessibility to the engine is achieved with the removal of ten bolts. Eight bolts removes the sheet metal and two catches take off the grille. Leaving uncompromising ability to service the engine to keep the 444 in top shape, the way it works best.

**Standard Equipment:**

3 point linkage; differential lock; molded, vertical acting suspension seat with backrest; constant running 545 rpm PTO; foot accelerator

**Optional Equipment:**

Belt pulley; down swept exhaust; self-sealing coupling; front and rear wheel weights; front end weights; single or double acting remote control attachment; swinging drawbar and frame; 3 point linkage drawbar; automatic hitch

*Photo Above: Pre-production photo of 444 tractor, showing the optional down swept exhaust*

**Specifications:**

Engine: AD-154 Diesel
Cylinders: 4
Type indirect injection
Piston displacement: 2.53 litre (154 cubic inch)
Power – Max gross rated: 42 hp (31 kW)
Rated governed speed: 2000rpm
Max Torque: 121 ft. lb. (164 Nm) @ 1400 rpm
Air Cleaner: Dry type with pre-cleaner
Cylinders, type: replaceable wet sleeves
Engine clutch Type: dual plate
Diameter: 11 inch (279mm) –transmission
9 inch (229mm) – PTO

**Transmission** (standard): eight forward and two reverse selected by gear lever and a high-low range lever

**PTO – constant running:** Rev/minute at governed engine speed
Standard: 545
Optional: 545/745
545 rpm single clutch (transmission drive)
PTO Power: 37hp (27.5 kW)

**Steering:**
Standard: Mechanical cam and lever
Optional: hydraulic power

**Brakes:**
Service: 6 ½ in (165mm) diameter totally enclosed outboard mounted discs, independent, foot operated
Parking: Over center, hand operated

**Electrical**
Battery: 12 volt
Safety starter: key start, clutch safety switch

**Instrumentation Gauges** – fuel, temperature, oil pressure, ammeter, tachometer

Turning Radius with brakes: 114in (2.90m)
Front wheel tread: 48-76in (1.22-1.93m)
Rear wheel tread: 53-76.5in (1.35-1.94m)

**Tyres:**

| | |
|---|---|
| Standard | – rear 13.6-28 |
| | – Front 6.00-16 |
| Optional | - rear 13.6-28; 14.9-28; 16.9-28; 14.9-24; 12.4-28 |
| | – Front 7.5-16 |

**Lighting**: two sealed beam headlights, rear working light

**Hydraulics and linkage**
Pump output: 6.3 gal/min (477ml/sec)
Pressure (max): 2300psi
3-point linkage: Category 1 and 2
Lift at link ends (approx.) 2600lb

**Capacities**

| | |
|---|---|
| Fuel: | 10.6 gal |
| Cooling system: | 2.5 gal |
| Engine oil: | 13 pints |
| Transmission case | 4.75 gal |
| Hydraulic Reservoir: | 2.5 gal |

Approx. operating weight: 4,840 lb. (2,195 kg)

**Dimensions**
Length: 118.25 in (3.0 m)
Wheelbase: 76.5 in (1.94m)

**Width, overall**
Front wheels, adj.: 59-87 in (1.50-2.21m)
Rear wheels, adj.: 64-88 in (1.63-2.24m)
Height, top of steering wheel: 59 in (1.50m)
Ground clearance (min under rear frame): 17.25in (438mm)

Tractor Speeds (at rated governed engine speed) on 13.6-28 tyres, mph (km/h)

| | |
|---|---|
| L1 | 1.46 (2.35) |
| L2 | 2.50 (4.02) |
| L3 | 3.56 (5.73) |
| L4 | 5.31 (8.54) |
| H1 | 4.26 (6.85) |
| H2 | 7.28 (11.71) |
| H3 | 10.37 (16.69) |
| H4 | 15.49 (24.92) |
| Low Reverse | 2.22 (3.57) |
| High Reverse | 6.46 (10.39) |

**Tractor Build Data**

Australian Tractor Test Number 79, June, 1973
Supplementary to test numbers 38 (A-414) and 47 (434)

Jan 1973 – Aug 1976

1,120 Units Built

| | |
|---|---|
| 1972/1973 | 178 |
| 1973/1974 | 238 |
| 1974/1975 | 444 |
| 1975/1976 | 160 |
| 1976/1977 | 100 |

For monthly production list, see Appendix D.

# Chapter 14
# 866

The Australian manufactured 866 is the more powerful complement to the 766 tractor. The 866 brings plentiful power with 16 forward speeds and comforts you need on this class of tractor.  To climb aboard this powerhouse of a tractor, it gives the feeling that no job is too big. Working on the hardest, most demanding tasks the International 866 has everything necessary for high productivity tractor work.

The work feature can only be used to the greatest effect if the operator is in total relaxed command. This tractor has an uncluttered operators deck, a fully adjustable, thickly cushioned, armchair style suspension seat, power steering and power brakes, easy shift gear and Torque Amplifier levers, full instrumentation, and hydraulic control levers right at hand.  Add an International harvester approved Gason weather cabin and protect yourself without bother from heat, rain, dust wind or cold.

The 866 open field tractor comes standard with a wide swinging roller drawbar, two speed independent power take off, double acting remote control hydraulics complete with double acting check valve and breakaway couplings. Add the optional three point linkage and this tractor is ready to take advantage of any situation.

The proven six cylinder D-358 provides 96 horsepower at 2,500rpm. Cylinder sleeves are replaceable, wet type. They dissipate combustion heat rapidly and are manufactured from alloy iron, centrifugally cast, honed and polished to increased wear resistance over other type sleeves. The precision machined, one-piece cylinder head is cast from special grey iron and secured to the cylinder block by 26 bolts that are specially arranged and equally spaced to ensure equal pressure on the head gasket and to prevent cylinder distortion.

Large intake and exhaust manifolds assure positive unrestricted air and fuel flows. Fuel filters are dual, full flow, spin-on, throw away type and filter the finest impurities. The dry type, dual stage air cleaner is highly efficient. It has an automatic dirt unloader, a washable element and a final built-in safety element. It works with the pre-cleaner baffle located in front of the radiator to keep dust out. An electric service indicator tells you when it needs cleaning.

This model also features the 'black stripe' on the late build models.

Shift lever linkage on the 866 lets you change gears with finger-tip effort. "H" pattern provides neutral between high and low gears. With the addition of the Torque Amplifier, eight basic forward speeds double to sixteen forward, eight reverse, and allows for power-shifting in-the-go. TA reduces speed by 22% while increasing torque up to 28% in each gear, with full engine braking in every gear. Use it through tough spots, or to slow down for turns. Get heavy loads started then shift up; keep PTO speed up while reducing ground speed; use TA to get the speed you want or work in it all day, it's built to take it.

**Ground Speeds (23.1 x 34 Tyres)**

| Range | Gear | Direct Drive | | Torque Amplifier | |
|---|---|---|---|---|---|
| | | Km/h | **Mph** | Km/h | **Mph** |
| Low | **1** | 3.09 | **1.92** | 2.42 | **1.50** |
| | **2** | 4.12 | **2.56** | 3.20 | **1.99** |
| | **3** | 7.15 | **4.44** | 5.57 | **3.46** |
| | **4** | 9.52 | **5.92** | 7.44 | **4.62** |
| High | **1** | 10.81 | **6.72** | 8.43 | **5.24** |
| | **2** | 14.39 | **8.93** | 11.20 | **6.97** |
| | **3** | 24.95 | **15.51** | 19.95 | **12.10** |
| | **4** | 33.20 | **20.69** | 26.00 | **16.14** |
| Reverse | **1** | 5.31 | **3.30** | 4.15 | **2.58** |
| | **2** | 7.06 | **4.39** | 5.50 | **3.42** |
| | **3** | 12.27 | **7.62** | 9.56 | **5.94** |
| | **4** | 16.35 | **10.17** | 12.78 | **7.93** |

Three separate hydraulic systems operate on the tractor. The first system operates the power steering, power brakes and constant pressure lubrication of transmission and differential. The second system powers the remote control cylinders and optional three-point hitch with lift capacity of up to 10,000 pounds (4,500 kg). The last system provides shifting and lubrication of the independent PTO. Each system only works when needed and rests in between times. The tubes are internal or shielded and all pumps are easily accessible for servicing. Dual rear outlets are standard on the 866.

Set and forget linkage, dual speed PTO, power shifting, side mounted batteries for easy servicing, rear operators platform (not on linkage models) for easy access are standard features.

*Above: 866 ready for final inspection*

*Left: 866 exiting the paint room*

**Specifications**

**Engine** D-358, 6 cylinder, Direct Injection
Piston displacement 358 cu.in. (5.9 liters)
Power – Max gross 121hp (90kW)
Rated Governed speed 2,500
Max Torque at 1600 rpm 265 lb.ft. (359Nm)

**Air Cleaner** Dry, with safety element and inbuilt pre-cleaner and additional pre-cleaner above radiator

**Engine Clutch**
Type 4 button "Dyna-Life"
Diameter 12 inch (305mm)

**Transmission**
Type Sliding Gear Torque Amplifier
Speeds 16 forward, 8 reverse
Final Drive Bull gear and pinion

**PTO Independent** Dual
Speeds 540 and 1,000
Engine rpm 2,300
PTO Power 96hp (72kW) at 2500rpm

**Steering Type** Hydrostatic Power

**Brakes**
Type Power Double Disc
Size 8 inch O.D. (203mm)
5 inch I.D. (127mm)

*Finishing touched are put on the 866*

**Electrical**
Batteries 2 – 6 volt
Safety Starter Key switch, clutch, safety switch, push button start

**Hydraulic System**
Reservoir Capacity 18.75 gal (85.25 liters)
Pump Type Gear
Pump Rated Output: 7.5 & 10 gpm
(568 & 757 ml/sec)
Pressure, max 2,000 psi (13,750 kPa)
Linkage, Optional Cat. 2
Lift at link ends 6,000 to 10,000 pounds
2,730 to 4,550 kg – depending on position

**Capacities**
Fuel 35 gal (159 liters)
Cooling System 5 gal (22.7 liters)
Engine Oil 2.75 gal (12.5 liters)
Transmission Case 18.75 gal (85.2 liters)

**Weight, Shipping** 10,790 pounds (4,900kg)

**Dimensions**
Length 159.4in (4.06m)
Wheelbase 104.8in (2.66m)
Width 91.0in (2.31m)
Height to steering wheel 86.95in (2.21m)

**Ground Clearance**
under drawbar pivot 17.1in (434mm)
under front axle 21.0in 533mm)

**Turning Radius**
with brakes 132in (3.35m)
without brakes 163in (4.14m)
Front Axle Adjustment 60 to 84 in 4 inch increments (1.52 to 2.13m in 100mm increments)
Rear Axle Adjustment: 68 to 106in (1.73 to 2.69m)
Roller drawbar Swing, without hitch 40 (1.02m)

**Tyres**

| | |
|---|---|
| Standard – Rear | 23.1x34 |
| – Front | 11.0x16 |
| Optional – Rear | 18.4x38 |
| | 18.x34 dual |
| | 23.1x30 dual |
| | 18.4x38 dual |

**Instrumentation** Gauges – Air cleaner indicator, fuel, ammeter, oil pressure, temperature, speedometer/tachometer, transmission oil pressure warning light

**Lighting** Two headlights grille mounted, rear flood light, combination stop and tail light, indirect light for instruments, optional front lights mounted in rear fenders

**Tractor Data**

Aug 1973 – May 1978

Australian Tractor Test number 86, April-May 1975

Price in 1976 $13,167

1,642 Units Built

| | |
|---|---|
| 1973/1974 | 173 |
| 1974/1975 | 237 |
| 1975/1976 | 389 |
| 1976/1977 | 473 |
| 1977/1978 | 356 |

For monthly production list, see Appendix D.

**Prototype Cabin of 866 tractor—**

notice the clips on top of the bonnet as well as the extended side panels and fins.

Taken at International Harvester Product Engineering Centre

# Chapter 15
# 976

The 976 is based on the proven '66' series International tractor design. It offer advantages traditionally associated with this design. Control systems are logically arranged, convenient and easy to use, power operated where you expect them. Heavy duty, power matched components are used throughout the tractor – six button, 14 inch Dyna-life clutch with long lasting cera-metallic facings; heavy duty final drive and axles; efficient engine air cleaning system; three pump hydraulic system.

The International 976 tractor offers you turbo-charged diesel power working through a sixteen forward, eight reverse speed transmission. Full power ground speeds for major farming tasks, the ground speeds have been selected to avoid the frustrating "one gear too low, one gear too high" situations to enable the farmer to get through their tasks quickly and efficiently.

This 358 cubic inch (5.9 litre), six cylinder diesel delivers 134 horsepower at the flywheel, 112 hp at the PTO, the muscle you need for harvesting, tillage, open field or row crop applications. The precision machined one-piece cylinder head is cast from special grey iron and secured to the cylinder block by twenty-six bolts. These bolts are arranged close to the cylinder and equally spaced to assure an even pressure on the cylinder head gasket and to prevent cylinder distortion. Rota-coils are fitted under the valve springs to give positive valve rotation, thus ensuring positive sealing and little wear. Pistons are cam-ground aluminum-silicon alloy with a stepped dome in the cylinder head.

In addition, the turbocharged power also offers excellent fuel economy over similarly rated but naturally aspirated engines. Turbo power uses the energy of exhaust gases to ram intake air into each cylinder.

More air means more complete combustion and more power per unit of fuel for better economy, more power per liter for less engine mass and size.

The wide power band of the 976 will allow you to shift up and throttle back, to the desired ground speed to conserve fuel. For example, during tillage operations, you may find that you have more power than you need at the optimum gear speed. All you do is shift to the next higher gear and throttle back to the desired ground speed. This way you maintain production, the engine isn't working as hard, resulting in conserving fuel.

Live type rear axles provide straight-through power to the rear wheels and are taper roller bearing mounted for long life and trouble-free service. Integral axle keys provide two driving surfaces, giving full power to the rear wheels at all times. Individual axle carriers bolt to the rear frame for easy maintenance and serviceability.

Full power hydraulic double disc brakes are self equalizing and self adjusting for wear, which provides safety and economical maintenance. The hydraulic powered multiple disc clutch has its own

hydraulic pump, with a twin shaft system that provides 540 and 1,000 rpm at an engine rpm of 2,300. Two stage clutch engagement gives a smooth PTO operation, allowing the PTO to start and stop gently to avoid overloads and strains.

A three pump hydraulic system provides dependable power to the various components. None of the pumps depend on the other for priming, pressure or volume. Only one type of fluid is required, as the pumps operate off a common reservoir. Filters with replaceable elements provide full-flow filtering. a swing-out hydraulic oil cooler, located in front of the engine radiator, allows for quick, easy inspection and cleaning – which keeps hydraulic fluid cool for long pump life and low maintenance and also reduces the frequency of oil changes.

At the business end of the 976 you'll find all you need to get your drawbar work underway and keep it under control. Two-speed power shift, independent PTO and dual rear hydraulic outlets are standard. Optional category 2 three point linkage has draft control lock-out for independent use of depth control irrespective of load. Remote control levers are located to the right of the seat for ease of accessibility.

Remote hydraulics have four positions; raise, lower, hold and float. The lever automatically returns to hold after release from either the raise or lower positions. Move the lever fully forward to provide the implement with float and it will remain in this position until manually repositioned. The hydraulics also has a lifting capacity of up to 10,000 pounds (4,550kg.)

With a wide swinging vertically adjustable drawbar with roller guide or three point linkage, wide variety of tyre sizes, including duals, there is a wide range of tread widths and linkage options to suit any requirement.

Side mounted twin batteries, rear operators platform (not on linkage models), adjustable front and rear wheels tread with optional dual rear wheels. Front end and rear wheel weights optional and rear floodlight and combination stop/tail lights are standard on all 976 tractors.

Vibration in the operators area has been reduced by the use of Isomount isolators – IH designed rubber mountings which absorb vibration. They have been installed in mounting the platform, steering column, gear selector support, fuel tank and sheet metal to the basic tractor. Take advantage of the 976 in all weather conditions with the International "Lo Sound" cab, specially designed for International '66' series tractors. It will keep you going after the sun goes down. Protected from the elements, productivity will be increased and you will be able to keep going, in comfort, making the hours worked much more pleasant.

**Ground Speeds (MPH) with 23.1 x 34 rear tyres**

| | Gear | Normal | TA |
|---|---|---|---|
| Lo Range | 1 | 2.0 | 1.6 |
| | 2 | 2.7 | 2.1 |
| | 3 | 4.6 | 3.6 |
| | 4 | 6.2 | 4.8 |
| Hi Range | 1 | 7.0 | 5.5 |
| | 2 | 9.3 | 7.3 |
| | 3 | 16.1 | 12.6 |
| | 4 | 21.5 | 16.8 |
| Reverse | 1 | 3.4 | 2.7 |
| | 2 | 4.6 | 3.6 |
| | 3 | 7.9 | 6.2 |
| | 4 | 10.6 | 8.3 |

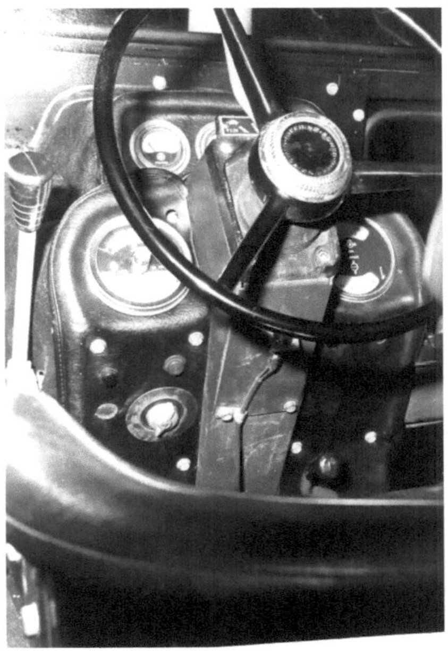

**Specifications**

**Engine**: International DT-358, 6-cylinder, direct injection

Piston Displacement: 3587cu.in. (5.9 litres)
Power (max gross) 134hp
Rated governed speed: 2,500
Max torque at 1,600 rpm: 327 lb.ft. (444 Nm)
Air Cleaner: Dry, with safety element and inbuilt pre-cleaner and additional pre-cleaner above radiator

**Engine clutch** type: 6 button "Dyna-Life", 14in (356mm) diameter

**Transmission:** Sliding gear Torque Amplifier
Speeds: 16 forward, 8 reverse
Final Drive: Bull gear and pinion

**PTO Independent:** Dual
Speeds: 540 and 1,000
Engine RPM, approx: 2,300
PTO Power at 2,500 rpm: 112hp (84kw)

**Steering:**
Hydrostatic Drive

**Brakes:**
Power Double Disc
Size: 8 in O.D. (203mm)
5 in I.D. (127mm)

**Electrical**
Batteries: 2 – 6 volt
Safety Starter: Key switch, Clutch, Safety Switch, Push Button Start

**Hydraulic System**
Reservoir Capacity: 18.75 gallons (85.25 litres)
Pump Type: Gear
Pump Rated output:
7 ½ & 10 gpm
(568 & 757 ml/sec)
Pressure (max): 2,000 psi (13,790kPa)
Linkage, optional: Cat. 2

**Capacities**

| | | |
|---|---|---|
| Fuel: | 35 gallons | (159 liters) |
| Cooling System: | 6 gallons | (27.3 liters) |
| Engine Oil: | 2.75 gal. | (12.5 liters) |
| Transmission Case: | 18.75 gal. | (85.2 liters) |

**Weight,** shipping: 11,000lbs (5,000kg)

**Dimensions**
Length: 159.4in (4.06m)
Wheelbase: 104.8in (2.66m)
Width: 97.6in (2.48m)
Height to steering wheel: 86.1in (2.19m)

**Ground Clearance**
Under drawbar pivot: 16.1in (408mm)
Under Front Axle: 21.7in (551mm)

**Turning Radius**
with brakes: 132in (3.34m)
without brakes: 163in (4.14m)

Front axle adjustment: 60-84inches, in 4-inch increments (1.52-2.13m in 100mm increments)
Rear axle adjustment: 68-106 inches (1.73-2.69m)
Roller drawbar swing without hitch: 40in (1.02m)

**Lighting:** Two headlights grille mounted, rear flood light, combination stop and tail light, indirect light for instruments, optional front lights mounted in rear fenders.

**Instrumentation:** Gauges – Air cleaner indicator, fuel, ammeter, oil pressure, temperature, speedometer – tachometer, transmission oil pressure warning light.

**Tyres**

| | |
|---|---|
| Standard – Rear | 23.1x34 |
| Standard – Front | 11.0x16 |
| Optional – Rear | 18.4x38 |
| | 23.1x34 |
| | 30.5x32 |
| | 18.4x34 dual |
| | 23.1x30 dual |

**Tractor Build Data**

Mar 1976 – July 1978

715 Units Built

1975/1976 88

1976/1977 344

1977/1978 245

For monthly production list, see Appendix D.

**Black Stripe Tractors**—The last of the series tractors that were build prior to the release of the next model. This was a common way to inexpensively freshen up the end of the model run in order to make sure the lot was empty and ready for the new models. The black stripe tractor models also have small changes that become standard on the next model. In this model, the 976 was replaced by the 986, and shows a front axle change.
The above photo taken in May 1977, is a pre-production 976 black stripe tractor.

Other Black Stripe Tractors that are Australian Built are the 766 and 866 models.

# Chapter 16
# 86 SERIES

The 86 Series tractors are the last tractor in the Australian built line-up. Built from 1979 until the factory closed in December 1982, the 786, 886, 986, 1086 and 1486 tractors were designed for the farmers comfort - not their employees. Standard Air conditioning, pressurized cabins to keep out dust and dirt, sound proofing and hydrostatic power steering were all for the owner-operators comfort.

IH muscle pulls you through with diesel engines to get you through your toughest farm work. The 786 tractor uses the D-310 diesel engine, proven performance and reputation in the field ensure that.

The 886 and 986 engines give you the muscle needed with harvesting, doing open field work or row crop applications. The 886 has a D-358 designed for long term economy. The Turbocharged DT-358 engine on the 986 makes it a prime tractor for any work. The 1086 with the DT-414 engine is an ideal tractor to cover primary and secondary tillage and seeding operations as well as a versatile performer in both broad acre and row crop farming. The most powerful Australian-built tractor in the range is the 1486 with the powerful turbocharged DT-436 which caters for the bigger farmer who wants to make the long days seem much shorter.

Easy monitoring of systems with a complete set of warning lights for radiator coolant level, oil pressure in transmission, pressure drop across the air cleaner and even a warning light for brake wear allow you to keep your eyes where they're needed most – on the crop.

The throttle lever is on the right side and the Torque Amplifier is on the left. Dimmer switch cuts down the glare from instrument lights, while key start with neutral start switch on clutch pedal, three position light switch, turn signals, cigarette lighter and horn are standard.

The cabin also features a new kind of quiet with thick rubber mounts between it and the tractor frame to suppress and isolate vibrations. Thick polyurethane foam and vinyl lining reduce sound in the cabin while the heavy-duty pressurizer system keeps out dust and dirt. The filter is self-purging with servicing from outside the cabin is simple.

Modular instrument cluster keeps you in the know with four gauges of plug-in type for easy servicing. In addition to the instrument cluster there is a tachometer/speedometer for the 786, 886 and 986 models. The 1086 and 1486 feature a digital finger touch readout—giving you and instant reading of ground speed, engine RPM, PTO speed and exhaust temperature.

Make a comfortable profit inside the control center. Standard air conditioning and a 6-way adjustable seat allows you to stay fresh and view your work comfortably.

Controls are logically placed for easy reach and electronic instruments give you easy monitoring.

Two door entry allows access from either side reducing the need to walk around a large implement to work on it. The doors have flush mounted handles outside and in, and the left hand door has a key lock.

Visibility in every direction gives you and unobstructed view either while you're driving, viewing the front wheels through the lower front windows or looking a big rear window to view hitch links and implement work. These are all benefits of the cabin control centre.

In every model of the 86 series, the basic transmission includes Hi-Lo range and eight speeds forward, four reverse. With the Tourque Amplifier, the transmission provides sixteen forward, eight reverse speeds, making sure there's the right speed for every job. The TA allows you to shift up or down on the go, without having to clutch or de-throttle.

The exclusive Dyna-Life clutch wears longer, dissipated heat better than conventional fiber clutches. It's cera-metallic pad facings last up to five times longer than others. Wet power brakes provide smooth sure stops under loads and help make tight fast turns. Multiple disc brakes are externally mounted and run in an oil bath for maximum heat dissipation and extended service life. Brakes automatically equalize and self adjust to compensate for wear, the brake pedals can be locked together and pedal faces are curved for comfortable foot operation.

*Above: Pre-production line-up of 986, 886 and 786. Notice muffler above bonnet on 786—in production all mufflers were under the bonnet. Below: view of engine.*

Electronic diff lock overcomes one wheel spin-out. It helps to maintain directional control and traction in soft and muddy footing. Post style pedal is located under the heel, close to the seat, to direct power equally to both drive wheels. Long life, multiple disc assembly is constantly lubricated by oil.

Unique IH pre-cleaner baffle is located behind the grille, re-directing airflow. The majority of dust and dirt particles fall before the air-stream enters the two stage dry type air filter. No pre cleaner servicing or cleaning is required and unlike other makes, there is no ugly pre-cleaner to obstruct your forward view.

An International diesel engine has earned its reputation through features such as: Bosh fuel injection pump—the single-plunger distributor type injection pump assures balanced fuel injection to all cylinders giving better combustion control for high efficiency and excellent fuel economy.

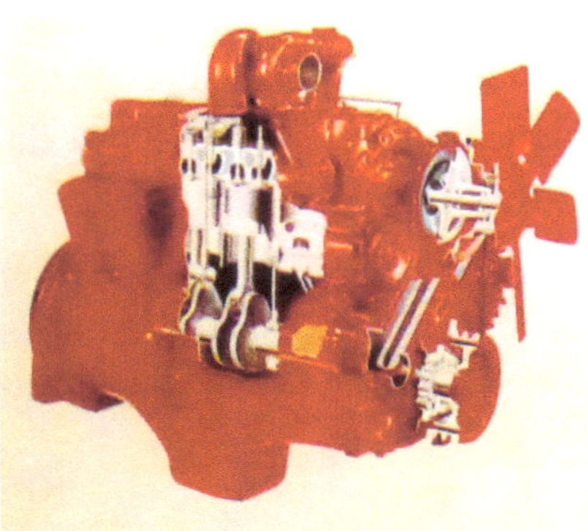

A quick glance at an IH diesel engine shows there's a minimum of external piping for a compact engine. Heavy Duty cylinder sleeves, crankshafts rods and bearings for extra strength and reliability are another part of the tough engine. Replaceable, wet-type sleeves and forged steel, induction hardened crankshaft with a vibration

damper on the front of the crankshaft as well as piston cooling jets on the turbocharged engines are other factors that make the diesel engine power.

The last aspect to provide the diesel power is the cooling and lubrication. The unique cooling design through the front cover gives the operator access to the lube, oil and coolant for a more compact design. The cooling circuit provides effective scrubbing action by spiraling coolant flow around the cylinder sleeves and upwards through the cylinder head water directors.

Gerotor oil pump on the front of the crankshaft delivers crankcase oil through the front cover to the oil cooler and to the spin on lube filters. These dual throw away full flow filters assure clean oil to all engine parts, filter only the oil needed to lubricate the engine, injection pump and turbo charger, providing extended engine life.

## Tractor Speeds

| Tractor Speeds km/h (mph) on Standard Rear Tyres: 23.1-34 786 | | | |
|---|---|---|---|
| Range | Gear | Direct Drive | Torque Amplifier |
| Low | 1 | 2.9 (1.8) | 2.3 (1.4) |
| | 2 | 3.8 (2.4) | 3.0 (1.9) |
| | 3 | 6.7 (4.1) | 5.2 (3.2) |
| | 4 | 8.9 (5.5) | 6.9 (4.3) |
| High | 1 | 10.1 (6.3) | 7.9 (4.9) |
| | 2 | 13.4 (8.3) | 10.5 (6.5) |
| | 3 | 23.3 (14.5) | 18.2 (11.3) |
| | 4 | 31.1 (19.3) | 24.2 (15.1) |
| Reverse | 1 | 5.0 (3.1) | 3.9 (2.4) |
| | 2 | 6.6 (4.1) | 5.1 (3.2) |
| | 3 | 11.5 (7.1) | 8.9 (5.6) |
| | 4 | 15.3 (9.5) | 11.9 (7.4) |

| Tractor Speeds km/h (mph) on Standard Rear Tyres: 23.1-34 886 and 986 | | | |
|---|---|---|---|
| Range | Gear | Direct Drive | Torque Amplifier |
| Low | 1 | 3.2 (2.0) | 2.5 (1.6) |
| | 2 | 4.3 (2.7) | 3.3 (2.1) |
| | 3 | 7.4 (4.6) | 5.8 (3.6) |
| | 4 | 9.9 (6.2) | 7.7 (4.8) |
| High | 1 | 11.3 (7.0) | 8.8 (5.5) |
| | 2 | 15.0 (9.3) | 11.7 (7.3) |
| | 3 | 26.0 (16.1) | 20.3 (12.6) |
| | 4 | 34.7 (21.5) | 27.1 (16.8) |
| Reverse | 1 | 5.5 (3.4) | 4.3 (2.7) |
| | 2 | 7.4 (4.6) | 5.7 (3.6) |
| | 3 | 12.8 (7.9) | 10.0 (6.2) |
| | 4 | 17.1 (10.6) | 13.3 (8.3) |

**Simple two-lever hitch control** is console mounted on the right for convenient operation. One lever sets draft and load control; the other raises or lowers the implement. If conditions change, nudge the lever forward to increase depth or rearward to decrease depth. A built in adjustable retarder valve controls rate of implement drop, regardless of its weight to protect against damage. A safety lock prevents accidental lowering of hitch.

| Tractor Speeds km/h (mph) on Standard Rear Tyres: 23.1-34 1086 | | | |
|---|---|---|---|
| Range | Gear | Direct Drive | Torque Amplifier |
| Low | 1 | 3.1 (1.9) | 2.4 (1.5) |
| | 2 | 4.1 (2.5) | 3.2 (2.0) |
| | 3 | 7.1 (4.4) | 5.5 (3.4) |
| | 4 | 9.5 (5.8) | 7.3 (4.6) |
| High | 1 | 10.7 (6.6) | 8.3 (5.2) |
| | 2 | 14.2 (8.8) | 11.1 (6.9) |
| | 3 | 24.6 (15.3) | 19.2 (11.9) |
| | 4 | 32.9 (20.4) | 25.7 (16.1) |
| Reverse | 1 | 5.2 (3.3) | 4.1 (2.5) |
| | 2 | 7.0 (4.3) | 5.4 (3.4) |
| | 3 | 12.1 (7.5) | 9.4 (5.9) |
| | 4 | 16.2 (10.0) | 12.6 (7.8) |

| Tractor Speeds km/h (mph) on Standard Rear Tyres: 23.1-34 1486 | | | |
|---|---|---|---|
| Range | Gear | Direct Drive | Torque Amplifier |
| Low | 1 | 3.3 (2.1) | 2.6 (1.5) |
| | 2 | 4.4 (2.7) | 3.4 (2.1) |
| | 3 | 7.6 (4.8) | 6.0 (3.7) |
| | 4 | 10.2 (6.3) | 8.0 (5.0) |
| High | 1 | 11.6 (7.2) | 9.0 (5.6) |
| | 2 | 15.4 (9.6) | 12.0 (7.5) |
| | 3 | 26.7 (16.6) | 20.8 (13.0) |
| | 4 | 35.6 (22.1) | 27.8 (17.3) |
| Reverse | 1 | 5.7 (3.5) | 4.4 (2.8) |
| | 2 | 7.6 (4.7) | 5.9 (3.7) |
| | 3 | 13.1 (8.2) | 10.2 (6.4) |
| | 4 | 17.5 (10.9) | 13.7 (8.5) |

## Specifications

| | 786 | 886 | 986 | 1086 | 1486 |
|---|---|---|---|---|---|
| **Engine**<br>International<br><br>**No. of Cylinders**<br>Type | D-310<br>Diesel<br>**6**<br>Direct Injection | D-358<br>Diesel<br>**6**<br>Direct Injection | DT-358 Turbo-charged Diesel<br>**6**<br>Direct Injection | DT-414 Turbo-charged Diesel<br>**6**<br>Direct Injection | DT-436 Turbo-charged Diesel<br>**6**<br>Direct Injection |
| **Piston Displacement liter (cu. in)**<br>Power (max Gross) hp<br>**Rated Governed Speed**<br>Air Cleaner | **5.08 (310)**<br>100<br>**2400**<br>Dry Type | **5.87 (358)**<br>121<br>**2500**<br>Dry Type | **5.87 (358)**<br>140<br>**2500**<br>Dry Type | **6.8 (414)**<br>148<br>**2400**<br>Dry Type | **7.1 (436)**<br>167<br>**2600**<br>Dry Type |
| **Engine Clutch**<br>Type<br>Diameter | "Dyna-Life"<br>305mm (12inches) | | "Dyna-Life"<br>356 mm (14inches) | | |
| **Transmission**<br>Type<br><br>**Speeds**<br>Final Drive | 8-speed Sliding Gear<br>Torque amplifier<br>**16 forward 8 reverse**<br>Bull Gear and Pinion | | | | |
| **PTO Independent Speeds**<br>Engine RPM for above | **Dual – 540 and 1000**<br>2100 | | | | **Single – 1000**<br>2300 |
| **PTO Power @rated engine speed** | **81hp** | **96hp** | **115hp** | **130hp** | **145hp** |
| Steering Type | Hydrostatic | | | | |
| **Hydraulic System**<br>(Rated output GPM)<br>**Chassis Functions**<br>Hitch & External out-puts<br>**PTO clutch pump**<br>Pressure Max (Hitch Pump) | **9.3**<br>10.3<br>**2.9**<br>2450 psi | **9.7**<br>10.8<br>**2.9**<br>2450psi | **9.7**<br>10.8<br>**2.9**<br>2450psi | **9.3**<br>10.3<br>**2.9**<br>2450psi | **9.4**<br>10.2<br>**2.9**<br>2450psi |
| **Brakes**<br>**Type**<br>Lining Diameter<br>Outside x Inside | **Hydraulic, full power wet multiple disc (5 per brake)**<br>8 inch x 5 inch | | **Hydraulic, full power wet multiple disc (6 per brake)**<br>8 inch x 5 inch | | |
| **Electrical**<br>Batteries<br>Safety Interlock | Two 12 volt low maintenance high capacity<br>Clutch Pedal or Park Lock Lever operated safety switch | | | | |
| **Capacities** gallons<br>Fuel<br>Cooling system<br>Engine Oil<br>Transmission Case | 47.6<br>5<br>2.75<br>18.75 | 47.6<br>5<br>2.75<br>18.75 | 63.5<br>6<br>2.75<br>18.75 | 63.5<br>5.2<br>4.2<br>18.75 | 63.5<br>5.7<br>4.2<br>22.4 |

Specifications continued

| | 786 | 886 | 986 | 1086 | 1486 |
|---|---|---|---|---|---|
| **Mass** Shipping Weight | 4,705kg (10,350lb) | 5,320kg (11,700lb) | 5,545kg (12,200lb) | 5,950kg (13,090lb) | 6,065kg (13,340lb) |
| **Length** (w/3pt linkage) | 3.96m (156in) | 4.06m (159in) | 4.06m (159in) | 4.27m (168in) | 4.27m (168in) |
| **Wheelbase** | 2.66m (105in) | 2.66m (105in) | 2.66m (105in) | 2.66m (105in) | 2.66m (105in) |
| **Width** | 2.59m (102in) | 2.59m (102in) | 2.59m (102in) | 2.59m (102in) | 2.59m (102in) |
| **Height** (over Cab) | 2.86m (112.4in) | 2.90m (114in) | 2.90m (114in) | 2.90m (114in) | 2.90m (114in) |
| **Ground Clearance** (under frame) | 343mm (14in) | 390mm (15in) | 390mm (15in) | 390mm (15in) | 390mm (15in) |
| **Turning Radius** With brakes Without brakes | 3.35m (132in) 4.14 m (163in) | 3.35m (132in) 4.14m (163in) | 3.35m (132in) 4.14m (163in) | 3.48m (136in) 4.06m (159in) | 3.48m (136in) 4.06m (159in) |
| **Front axle adjustment** | 1.52 to 2.13 m (60 to 84 in) | 1.52 to 2.13 m (60 to 84 in) | 1.52 to 2.13 m (60 to 84 in) | 1.54 to 2.2m (61 to 85in) | 1.54 to 2.2m (61 to 85in) |
| **Rear axle adjustment** with Standard tyres, single | 1.7 to 2.8 m (68 to 113 in) | 1.7 to 2.8 m (68 to 113 in) | 1.7 to 2.8 m (68 to 113 in) | 1.7 to 2.8 m (68 to 113 in) | 1.7 to 2.8 m (68 to 113 in) |
| **Tyres** Standard Rear Standard Front | 23.1 x 30 13.0 x 16 | 23.1 x 34 11.0 x 16 | 23.1 x 34 11.0 x 16 | 24.5 x 32 11.0 x 16 | 24.5 x 32 11.0 x 16 |

**Tractor Build Data**

**86 Series Tractors**

**786** Aug 1978 – Dec 1980

997 Units Built

**886** June 1978 – Feb 1981

826 Units Built

**986** Sept 1978 – Jan 1981

492 Units Built

**1086** May 1979 – Nov 1980

296 Units Built

**1486** July 1979 – July 1980

70 Units Built

For monthly production list, see Appendix D.

***View of digital dash in the 1086 and 1486.***

**What is the SLE?**

*This tractor was made in-between the change over to the B Series tractors. The decals on the side of the tractor and on the front fender read SLE, and appears to be an open station 786, according to the serial number listed.*

*The Engine number is also from the 786 lineup, but dates at a later stage— after beginning of production of the B Series tractors.*

*These Company photos are dated June 1981 and show the SLE in a cane field. Was this tractor meant to be aimed towards the Hi-Clear needs?*

# Chapter 17
# 86 B SERIES

**86 B Series Tractors**

With the introduction of the 86 series tractors in 1979, Australia learned that comfort, performance and convenience could be in a tractor. Gone were the days of hot dusty conditions with no respite from the harsh Australian sun. Gone were the days of not quite fast enough or just a bit slower. Gone were the days of heavy steering, bone jarring rides and anything else that 'just had to do.' With the introduction of these luxuries, there is always room for improvement. The 86 series tractors brought the operator comfort and the new B Series, introduced in 1981, improved upon this beginning and added a host of new changes that has set the standard for tractors today.

Visual changes to the B Series tractor included the following: front engine panel changed from white to red; red stripe added to top of pinstriping on sides of tractor; Cabin features red and black scheme instead of white; door panels on inside of cabin are now red with a pocket—previously the door was black trim without storage pocket; simple pull loop replaced with proper door closing handle; Front weights now in black; top grille insert was silver, now painted black; large side mirrors; removable toolbox and a better opening rear window complete the visual changes to the tractor.

**86 Series features are:**

- High torque diesel engine, designed for efficient reliable power.
- Heavy duty power shifting torque amplifier transmission gives 16 forward speeds
- Hydraulic power assisted Dyna-life clutch for low pedal effort and long life
- Self adjusting, automatically equalizing multi-disc wet power brakes
- Electro hydraulic diff lock (3 point linkage models)
- Independent lift, PTO and auxiliary hydraulic pumps with full flow suction line filtration
- Large capacity fuel tanks for extended field operation
- A control centre that sets new standards in comfort and styling

**New Features for the 86 B Series.**

- Seat suspension for improved ride over high speed, rough conditions.
- CAT III 3-point linkage available for 786 & 886 Series B models for heavy duty operation.
- CAM sway limiters for positive side sway control of 3-point linkage
- 200 hour extended engine oil change period
- Contoured grip type gearshift knobs plus new floating high-low shift rails.
- Tungsten halogen 50 watt sealed beam head and front flood lamps increases forward night vision.
- One piece front axle bolster with parallel stayrod pivot pin
- Nodular cast iron front axle centre section

- Customer support pack—an industry first. Kit includes filters for first 500 hours operation, drive belts and electrical connector.
- Color coded hydraulic valve lever knobs and breakaway couplings
- Addition of 2 sun visors to control centre
- Addition of 2 external "west-coast" type mirrors—standard on 986, 1086, 1486 and optional on 786 and 886—allows wide vision to the rear
- Door operated courtesy switched for dome light
- Burgundy color interior trim
- Pockets in each door and pull handles
- Steering wheel cap trim
- Toolbox is removable and lockable

- Nodular cast iron front ballast weight mounting bracket and simplified cast weight with retaining bar and CAM type side clamping makes changing weights easy.

***Clockwise: Customer support pack; Cabin door showing the burgundy color trim with pockets and pull handle; large opening rear and side windows; Westcoast style mirrors allow for panoramic viewing.***

**A one piece front axle**—and bolster replaces several parts and features a stayrod pin. This design will maintain its rigid characteristics. The cast iron front axle centre section complements the bolster and allows for easy adjustment of the front wheel tread.

Front wheel assist is an optional extra for the B series tractors. It also is a new feature to the 86 Series B tractors. This allows power normally lost through wheel slippage and rolling resistance to provide four wheel traction for all weather conditions. Seventeen degrees oscillation each way allows for usage on uneven ground or working in furrows.

***Right: 1086 with Optional front wheel assist axle.***

***Getting the new look 86 Series "Just Right"***

*Pre-Production photos of the 86 Series B Tractors.*

*Photos date 10 July, 1980.*

*Note the different cabin colors and decal schemes on each tractor.*

*Decal in Top photo is of a 1486 Black Stripe with Titan II on bottom of the side engine plate. White cabin with black side posts and white grill with white and black light surround completes the look.*

*Middle Photo is 1086B with white cabin, black side posts and a black grille with black and white light surround.*

*Bottom Photo is of a 1486 without a decal to differentiate that it is a 'B Series.' It features a red cabin, black side posts, black grille and light surround with black weights and weight bar.*

*Bottom Right Photo is the Final product. The Series B decal is placed prominently on the door, with a red cabin, black side posts, black and white light surround with a white grille. It also has black weights and a red weight bar.*

**Front Tyres**

14.9 x 24 for operation with these rear tyres.

| | |
|---|---|
| 18.4 x 38 | 20.8 x 38 |
| 23.1 x 34 | 24.5 x 32 |

**Special**
14.9 x 28 front tyres matched with 20.8 x 38 rear tyres for Hi-Clearance

**Speeds and Specifications for the 86 B Series are the same as the 86 Series tractors.**

**Optional extras for the B Series are:**

4WD Assist Front Axle;
Tread settings:
1981mm (78 inches) or 2082mm (82 inches)
1727mm (68inches) or 2336mm (92 inches)

**Tractor Build Data**

Sold from March 1981 until November 1982.
Monthly build data in Appendix.
*Note that while some models were truly finished being built, (as noted) a couple of tractor models were assembled after the official close in December 1982. These were most likely built using parts that were already designated for the assembly line.*

**786B**

Mar 1981 – Aug 1982
441 Units Built

**886B**

Feb 1981 – Nov 1982
432 Units Built

**986B**

Feb 1981 – Sept 1982
328 Units Built

**1086B**

Feb 1981 – Nov 1982
164 Units Built

**1486B**

Feb 1981 – Oct 1982
64 Units Built

*Before cabins can be approved for safety, they must be tested. The following eight photos are an insight to the testing process. They show the impact and crush test of an 86 series tractor. These tests were conducted at the State Research Facilities in Werribee, Victoria.*

*Above: Preparing the tractor for testing*

*Right: First impact test, cabin rear*

*Rear Impact*

*Side impact test set-up*

*Setting up for forward impact*

*Forward impact test in action*

*Crush Test of cabin, forward view*

*Crush test of cabin, rear view*

# Chapter 18
# ACREMASTER

The Acremaster tractors were sourced from Phillips in Western Australia by International Harvester of Australia to fill the need for a large, broad acre tractor during 1984 and 1985. The only changes made to the tractors when they went to the IH brand was the paint and decals.

Designed for the varied operating and climatic conditions existing from northern Qld to WA, the Acremaster has proved that Australian know-how and service backup is a distinct advantage in maintaining on-farm profitability.

The Acremaster range covers all broad acre operations. "Overbuilt" specifications and simplicity of design provide the best in 4wd performance, with the benefits of easy maintenance and longer operational life that keep downtime to a minimum.

Hand built quality; operator comfort and one of the quietest cabs in the industry make the tropical cabin a pleasure to work in. Exceptional visibility, ergonomic design, and spacious two seater interior cut down operator fatigue, while the large roof overhang, outward sloping tinted glass and commanding driving position keep the sun glare to a minimum. A super efficient air condition system copes with the worst tropical climate with ease and the 'Thermalmaster' combination refrigerator and oven help shorten those long tractor hours and ease the strain on the cook.

*Right: combination fridge and oven, complete with meal trays and upholstered passenger seat lid.*

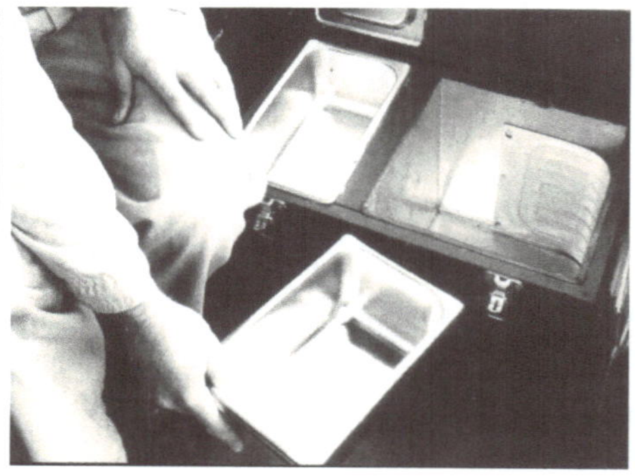

A model for every operation, with the choice of 9-speed manual or 12 speed full power shift transmission to maximise tractor performance over varied ground and operation conditions. Acremaster tractors are designed to perform under the toughest conditions, with up to 44" chassis oscillation to maintain constant 4wd over the roughest ground conditions.

Acremasters have a proven track record under the worst conditions in Australia and are designed to go on performing around the clock to keep your farm operation viable.

Meticulously built by a team of young Australian craftsmen devoted to the concept of quality, and serviced by highly trained staff whose first priority is to keep the wheels turning, the Acremaster is an asset to the Australian agricultural industry.

*Left: Modular, simple construction with easy accessibility for servicing.*

Features of the Acremaster are:

- 9 Models –from 210hp to 525hp
- Unique three compartment, modular steel plate chassis to cope with heavy duty operations
- Oscillating rear axle pivots 44 degrees, reduces suspension shock transmitted through the chassis to provide ride comfort
- Little extras—an air compressor system fro power tool operation, an electrical isolator for child safety and electrical power drain prevention, easy to adjust 24volt floodlights and spotlights for night work, large dust free tool boxes, 24 hour fuel range, two seats.
- Effortless performance and economy from large displacement under stressed engines, which allows the flexibility of applications for Australian farmers.
- High efficiency air conditioner
- Tinted windows slope 4 degrees to prevent glare and dust build-up

**Specifications:**

**Fuel Capacity:**

**A210, A250, A310, A310PS, A330, A330PS—** Approximately 1,016 liters (200 gallons)

**A400, A400PS—** 1,520 liters (300 gallons)

**A520PS—** 2,030 liters (400 gallons)

**Electrical: All models—** 80amp. 24 volt Electro-dyne alternator , brushless, solid state regulator, quartz halogen lights

Late Acremaster tractors featured a Cummins engine rather than the Mercedes-Benz Engines.

**Acremaster Series Engine Specifications**

| Models | Power-Engine | Engine Model | Piston Displacement (liters) | Aspiration | Number of Cylinders | Stroke |
|---|---|---|---|---|---|---|
| **A210** | 210 hp | MB OM421 | 10.9 | Natural | V6 | 142mm |
| **A250** | 249hp | MB OM402 | 12.76 | Natural | V8 | 130mm |
| **A310** | 308hp | MB OM403 | 15.95 | Natural | V10 | 130mm |
| **A310PS** | 308hp | MB OM403 | 15.95 | Natural | V10 | 130mm |
| **A330** | 322hp | MB OM422A | 14.62 | Twin Turbo-chargers | V8 | 142mm |
| **A330PS** | 322hp | MB OM422A | 14.62 | Twin Turbo-chargers | V8 | 142mm |
| **A400** | 400hp | MB OM404 | 20.9 | Natural | V12 | 142mm |
| **A400PS** | 400hp | MB OM404 | 20.9 | Natural | V12 | 142mm |
| **A520PS** | 515hp | MB OM404A | 20.9 | Twin Turbo-chargers | V12 | 142mm |

***PS** stands for Power Shift on the above models.*

<table>
<tr><th>Models</th><th>Transmission</th><th>Clutch</th><th>Axles</th><th>Hydraulics</th><th>Chassis</th></tr>
<tr><td>A210</td><td rowspan="3">Acremaster 9-speed constant mesh, pressure lubricated, free standing gear-box</td><td rowspan="3">387mm twin plate ceramic Spicer clutch</td><td rowspan="2">Acremaster large capacity outboard plane-tary hub with spiral bevel dif-ferential</td><td>Open center systems, 2 re-mote outlets, individual gear pumps</td><td rowspan="2">Heavy Duty 12mm (1/2 inch) plate</td></tr>
<tr><td>A250</td><td>Open center systems, 3 re-mote outlets, individual gear pumps</td></tr>
<tr><td>A310</td><td rowspan="7">Rockwell large capacity out-board planetary hub with hypoid differential</td><td rowspan="6">Open center systems, 4 pairs remote outlets, individual gear pumps</td><td rowspan="4">Heavy Duty 16mm (5/8 inch) plate</td></tr>
<tr><td>A310PS</td><td>Twin disc, full powershift 12 speed computer shift (1407 se-ries)</td><td>N/A</td></tr>
<tr><td>A330</td><td>Acremaster 9-speed constant mesh, pressure lubricated, free standing gear-box</td><td>387mm twin plate ceramic Spicer clutch</td></tr>
<tr><td>A330PS</td><td>Twin disc, full powershift 12 speed computer shift (1407 se-ries)</td><td>N/A</td></tr>
<tr><td>A400</td><td>Acremaster 9-speed constant mesh, pressure lubricated, free standing gear-box</td><td>387mm twin plate ceramic Spicer clutch</td><td rowspan="2">Heavy Duty 25mm (1 inch) plate</td></tr>
<tr><td>A400PS</td><td>Twin disc, full powershift 12 speed computer shift (1407 se-ries)</td><td>N/A</td></tr>
<tr><td>A520PS</td><td>Twin disc, full powershift 12 speed computer shift (2616 se-ries)</td><td>Lock-up torque converter</td><td>Open centre system with 4 pairs remote outlets, plus 2 hydromotor PTO drive op-tion, individual gear pumps</td><td>Heavy Duty 38mm (1 1/2 inch) plate</td></tr>
</table>

***Typical work role of the A400.***

<table>
<tr><th>Model</th><th>Standard Tyres</th><th>Shipping weight</th><th>Operating Weight</th></tr>
<tr><td>A210</td><td>18.4 x 34 Duals or<br>30.5 x 32 singles</td><td>9 Tons</td><td>11.5 Tons</td></tr>
<tr><td>A250</td><td>23.1 x 30 Duals or<br>30.5 x 32 Singles</td><td>9.7 Tons</td><td>12 Tons</td></tr>
<tr><td>A310</td><td rowspan="4">24.5 x 32 Duals</td><td rowspan="4">11.5 Tons</td><td rowspan="2">14.8 Tons</td></tr>
<tr><td>A310PS</td></tr>
<tr><td>A330</td><td rowspan="2">15 Tons</td></tr>
<tr><td>A330PS</td></tr>
<tr><td>A400</td><td rowspan="2">30.5 x 32 Inners<br>and<br>24.5 x 32 outers</td><td rowspan="2">15.8 Tons</td><td rowspan="2">20 Tons</td></tr>
<tr><td>A400PS</td></tr>
<tr><td>A520PS</td><td>30.5 x 30.5 Duals</td><td>20 Tons</td><td>27 Tons</td></tr>
</table>

**Tractor Sales Data:**

**1984-1985, number sold**

| | |
|---|---|
| **A210** | 16 |
| **A250** | 9 |
| **A310** | 8 |
| **A330** | 20 |
| **A400** | 4 |

*Large but compact engines are easy to work on.*

## APPENDIX A

### Geelong Built PTO Headers

| PTO Driven | YEARS BUILT | TOTAL NUMBER BUILT | CUT WIDTHS |
|---|---|---|---|
| GL200 | 1947-1957 | 6715 | 10 foot |
| A8-1 | 1959-1961 | 3134 | 12 or 14 foot |
| A8-4 | 1962-1973 | 4920 | 12 or 14 foot |
| 8-6 | 1968-1976 | 1835 | 14 or 16 foot |
| 710 | 1977-1979 | 1082 | 14, 16 or 18 foot |
| 726 | 1980-1982 | 725 | 14, 16 or 18 foot |

**Total Production of PTO Driven Headers 18,411**

| GL200 | | A8-1 | | A8-4 | | 8-6 | | 710 | | 726 | |
|---|---|---|---|---|---|---|---|---|---|---|---|
| Year Built | Number Built | Year Built | Number Built | Year Built | Number Built | Year Built | Number Built | Year Built | Number Built | Year Built | Number Built |
| 1947 | 1 | 1959 | 926 | 1962 | 800 | 1968 | 20 | 1977 | 400 | 1980 | 274 |
| 1948 | 100 | 1960 | 1406 | 1963 | 1000 | 1969 | 275 | 1978 | 400 | 1981 | 319 |
| 1949 | 502 | 1961 | 800 | 1964 | 1000 | 1970 | 60 | 1979 | 280 | 1982 | 129 |
| 1950 | 601 | | | 1965 | 800 | 1971 | 100 | | | | |
| 1951 | 600 | | | 1966 | 350 | 1972 | 150 | | | | |
| 1952 | 600 | | | 1967 | 420 | 1973 | 200 | | | | |
| 1953 | 1152 | | | 1968 | 300 | 1974 | 350 | | | | |
| 1954 | 1460 | | | 1969 | 150 | 1975 | 350 | | | | |
| 1955 | 799 | | | 1970 | - | 1976 | 330 | | | | |
| 1956 | 800 | | | 1971 | - | | | | | | |
| 1957 | 100 | | | 1972 | 79 | | | | | | |
| | | | | 1973 | 21 | | | | | | |

*Farmall M Tractor with GL-200 Header*

## Geelong Built Self-Propelled Headers

| Self-Propelled | YEARS BUILT | TOTAL NUMBER BUILT | CUTTING WIDTH |
|---|---|---|---|
| A8-2 | 1959-1961 | 2189 | 10 foot |
| A8-3 | 1961-1962 | 244 | 10 foot |
| 8-5 | 1966-1973 | 1219 | 14 or 16 ft comb; 12, 14, or 16 ft open |
| 711 | 1973-1978 | 2326 | 16, 18 or 20 foot comb; 16 ft open |
| 725 | 1979-1981 | 1310 | 16, 18, or 20 foot comb; 16 ft open |

*Last 33 '725' were renumbered, (s/n 1500-1533) updated and sold as Mark II*

**Total production of Self-propelled Headers 7,288**

| **A8-2** | | **A8-3** | | **8-5** | | **711** | | **725** | |
|---|---|---|---|---|---|---|---|---|---|
| YEAR BUILT | NUMBER BUILT | YEAR BUILT | NUMBER BUILT | YEAR BUILT | NUMBER BUILT | YEAR BUILT | NUMBER BUILT | YEAR BUILT | NUMBER BUILT |
| 1959 | 12 | 1961 | 130 | 1966 | 20 | 1973 | 196 | 1979 | 775 |
| 1960 | 200 | 1962 | 114 | 1967 | 344 | 1974 | 403 | 1980 | 375 |
| 1961 | 6 | | | 1968 | 406 | 1975 | 502 | 1981 | 160 |
| | | | | 1969 | 250 | 1976 | 500 | | |
| | | | | 1970 | - | 1977 | 525 | | |
| | | | | 1971 | - | 1978 | 200 | | |
| | | | | 1972 | 152 | | | | |
| | | | | 1973 | 47 | | | | |

***A8-3 Self-propelled Header; only model to use levers for steering instead of the traditional wheel.***

## APPENDIX B

## Geelong Built Implements

In the early days, implements started with the prefix 'GL' and later 'A'. During the 1960s, the prefix was dropped.

**Early Implements 1939-1946**
***(The basic line that was built during WWII)***

| | |
|---|---|
| GL-9 | Horse Drawn Mower |
| GL-9A | Tractor Disc Harrow |
| GL-42 | Little Genius Plow |
| Gl-43 | Little Genius Plow |
| GL-90 | Tractor Disc Cultivator Plow |
| GL-104 | Hay Rake |
| GL-125 | Fertilizer Distributor; 7 or 9 foot; horse or tractor |
| GL-130 | Cultivator Drill |
| GL-150 | Tool Bar |
| GL-300 | Gas Producer Unit |

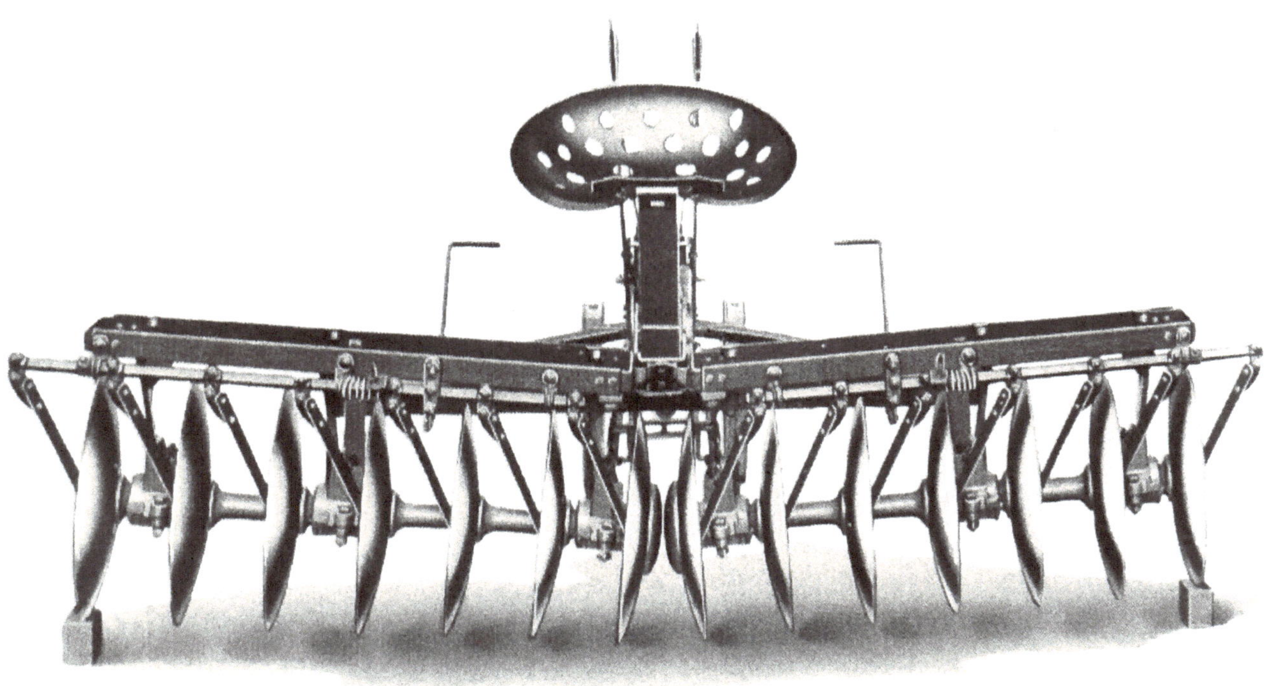

## Geelong Built Implements 1946-1982

A Comprehensive listing of Implements made at Geelong Works. Most of the Construction implements were made at Port Melbourne Works. These are not listed. Implements made outside of Geelong Works are not listed as well.

| | |
|---|---|
| A-68D-4 | Combination Carrier with A30-1 Toolbar (crawler tractors) |
| A-69D-4 | Combination Carrier and Light Blade (crawler tractors) |
| A1-2 | Disc Plow – 12, 14, 16 or 20 disc<br>Seeder Attachment Available |
| A1-35 | Reversible Plow |
| A1-41 | Plow—Disc or Moldboard; 3-pt linkage; two, three, or four furrow |
| A1-54 | Plow—Disc or Moldboard; 3-5 furrow |
| 1-6 | Stump Jump Plow 14, 18 or 22 disc |
| A2-1 | Cultivator – Parallel lift type, mid-mounded, rear section for 3-pt available |
| A2-2 | Cultivator; 4-row |
| A2-3 | Cane Cultivator; Two-row |
| A2-6 | Cultivator; 4-row |
| 2-11 | Scarifier |
| A21-1 | Side Delivery Rake 8 ½ -foot raking width |
| A21-2 | Buck Rake 9-foot wide, 12 tyne tubular |
| A21-3 | Side Delivery Rake; Ground Driven |
| A3-1 | Una-Beam Disc Harrow |
| 3-2 | Bush and Bog Harrow; 16 or 20 disc |
| 3-3 | Harrow |
| 3-4 | Harrow |
| 3-5 | Harrow |
| A30-1 | Crawler Tractor Toolbar |
| A30-2 | Row-Crop Toolbar Two- row; 5, 6, 7, 8, 9, 10, 12 or 14' width |
| A30-3 | Planter Toolbar; 12' bar |
| A30-4 | Open Field Toolbar; 5, 6, 7, 8, 9, 10, 12 or 14' width |
| 30-5 | Planter Toolbar; 12' bar |
| 30-6 | Row Crop Toolbar; Single Row; 5, 6, 7, 8, 9, 10, 12 or 14' width |
| 30-7 | Row Crop Toolbar; Two Row; 5, 6, 7, 8, 9, 10, 12 or 14' width |
| 30-8 | Open Field Toolbar 5, 6, 7, 8, 9, 10, 12 or 14' width |
| 31-1 | Liquid Fertilizer Distributor; 70-gallon |
| A39-1 | Carry Lift Platform |
| A4-1 | Rotary Cutter; 6-foot cut |
| A4-3 | Balanced Head Mower |
| A4-4 | Balanced Head Mower |
| A4-5 | Pitman Mower |
| A4-6 | Rotary Mower |
| A4-7 | Rotary Slasher |
| A4-8 | Rotary hay Mower |
| A4-9 | Rotary Shredder |
| A40-1 | Superloader |
| A41-1 | Rotary Mower |

| | |
|---|---|
| A41-1 | Mobile Crane |
| A41-1 | Disc Plough |
| A42-1 | Hay Conditioner; 7-foot width |
| 45 | Vibrashank |
| 511 | Drill; Replaces 6-2 |
| A6-1 | Cultivator Drill;<br>16, 20 or 24-row 7-inch spacing |
| 6-2 | Drill;<br>Finished in 1973, Replaced by 511 |
| 740 | Scarifier |
| A-84-HB | Blade; Angling and Tilting |
| AUD-264 | Diesel Power Unit |
| GL-9A | Disc Harrow - 5, 6, 7, 8 or 10-foot |
| GL-13 | Disc Harrow - 6 ½ foot |
| GL-14 | Disc Drill – 5 ¼ or 6-foot |
| GL-21M | Harrow – 8 to 10 foot pasture type<br>and 12 to 15 foot |
| GL-23M | Harrow- 1 to 10 section Diamond<br>smoothing; 1 to 15 section |
| GL-25M | Harrow- 1 to 10 section<br>Stump Jump type; 1 to 15 section |
| GL-25VO | Tractor Mower; 5, 6 or 7 foot |
| GL-32 | Scarifier; 16 or 18 tyne |
| GL-55 | Automatic pickup baler |
| GL-71 | Tractor Disc Plow |
| GL-100 | Side delivery rake |
| GL-115 | Hay Press, Belt Pulley |
| GL-116 | Hay Press, PTO |
| Gl-132 | Cultivator- Drill;<br>16-row or 20 row (planter) |
| GL-139 | Potato Planter Single row – horse;<br>Double row - Tractor |
| GL-142 | Cane Cultivator; Two-row |
| GL-143 | Grubber; Two-row |
| GL-163 | Heavy Duty tool bar; BTD-6 only |
| GL-164 | Heavy Duty tool bar; BTD-6 only |
| GL-187 | Cultivator; Two-row |
| GL-222 | Cultivator; Spring Tyne |
| GL-223 | Mower; 5 or 6-foot |
| GL-224 | Ripper |
| GL-225 | Post Hole Digger |
| GL-226 | Blade; light rear blade; 6 or 8-foot |
| GL-227 | Toolbar- Open Field; 7 or 8-foot |
| GL-227 | Cultivator |
| GL-229 | Tool Bar- Orchard and Vineyard;<br>6 or 7-foot |
| GL-230 | Tool Bar- Row-crop;<br>Two 25-inch sections;<br>6 to 10-foot – open field |
| GL-240 | Loader; 10 cubic foot capacity |
| GL-250 | Feed Grinder with bagging elevator;<br>belt pulley operation |
| GL-251 | Hammer Mill; Belt pulley operation |
| GL-260 | Milking Machine |
| GL-330 | Scoop ½ -yard |
| GL-331 | Scoop ⅔ -yard |
| GL-332 | Scoop ¾ -yard |
| GL-333 | Scoop 1 -yard |
| GL-325 | Offset Disc Harrow; 5 ¼ or 6 foot |

| | |
|---|---|
| GL-326 | Carry-lift Platform |
| GL-327 | Harrow Pasture type; 10, 12 or 15 ft |
| GL-334 | Ripper |
| GL-350 | Tool bar; - Orchard, Vineyard and Market garden work; 60, 70 and 87-inch lengths; double channel type |
| GL-351 | Moldboard Plow; Three furrow |
| GL-352 | Cultivator; Rigid Tyne |
| GL-355 | Cultivator; Spring Tyne |

**Defender Refrigerators** – Made for International Harvester Australia by Electrolux Australia as early as 1946

Model Defender – 5 ½ cubic foot; 10 square feet of shelf area, operates on Kerosene

Model X 701 – "Country Style" 7 cubic foot, 13 square feet of shelf area, operates on Kerosene or Electricity

Model XT-70 – "Selectra Style" 7 cubic foot, 13 square feet of shelf area, sealed unit, operates on electricity

# APPENDIX C

## Geelong Tractor Production 1948-1959

| Period | AW-6 | AM | SUPER AW-6 | SUPER AWD-6 | AOS6 | SUPER AM | SUPER AMD | AW7K | AW7D | AM7K | AM7D | TOTAL | PROGRESSIVE TOTAL |
|---|---|---|---|---|---|---|---|---|---|---|---|---|---|
| 1947/48 | 266 | | | | | | | | | | | 266 | 266 |
| 1948/49 | 738 | 200 | | | | | | | | | | 938 | 1,204 |
| 1949/50 | 1,316 | 700 | | | | | | | | | | 2,016 | 3,220 |
| 1950/51 | 1,951 | 1,000 | | | | | | | | | | 2,951 | 6,171 |
| 1951/52 | 2,250 | 1,001 | | | | | | | | | | 3,251 | 9,422 |
| 1952/53 | 888 | 440 | | | 6 | | | | | | | 1,334 | 10,756 |
| 1953/54 | | | 1,137 | 745 | 868 | 130 | 120 | | | | | 3,000 | 13,756 |
| 1954/55 | | | 663 | 1,255 | 800 | 82 | 200 | | | | | 3,000 | 16,756 |
| 1955/56 | | | 250 | 924 | 150 | 51 | 92 | | | | | 1,467 | 18,223 |
| 1956/57 | | | 148 | 679 | 75 | 1 | 52 | 1 | 106 | | | 1,062 | 19,285 |
| 1957/58 | | | 59 | 65 | 7 | | 1 | 46 | 871 | 8 | 78 | 1,135 | 20,420 |
| 1958/59 | | | | | 14 | 1 | 7 | 140 | 1,244 | 7 | 65 | 1,478 | 21,898 |
| **TOTAL** | **7,409** | **3,341** | **2,257** | **3,668** | **1,920** | **265** | **472** | **187** | **2,221** | **15** | **143** | **21,898** | |

# Geelong Tractor Production 1960-1982

| Period | A-554 | A-514 | A414 | 564 | 434 | 686/ 756 | 696/ 766 | 444 | 866 | 976 | 786 | 786 B SERIES | 886 | 886 B SERIES | 986 | 986 B SERIES | 1086 | 1086 B SERIES | 1486 | 1486 B SERIES | TOTAL | PROGRESS TOTAL |
|---|---|---|---|---|---|---|---|---|---|---|---|---|---|---|---|---|---|---|---|---|---|---|
| 1960/61 | 1 | | | | | | | | | | | | | | | | | | | | **1** | 1 |
| 1961/62 | 2,333 | | | | | | | | | | | | | | | | | | | | **2,333** | 2,334 |
| 1962/63 | 1,983 | 150 | | | | | | | | | | | | | | | | | | | **2,133** | 4,467 |
| 1963/64 | 1,730 | | 1,500 | | | | | | | | | | | | | | | | | | **3,230** | 7,697 |
| 1964/65 | 1,856 | | 2,321 | | | | | | | | | | | | | | | | | | **4,177** | 11,874 |
| 1965/66 | 815 | | 1,161 | | | | | | | | | | | | | | | | | | **1,976** | 13,850 |
| 1966/67 | 85 | | 432 | 802 | 776 | | | | | | | | | | | | | | | | **2,095** | 15,945 |
| 1967/68 | | | | 672 | 1,251 | | | | | | | | | | | | | | | | **1,923** | 17,868 |
| 1968/69 | | | | 444 | 816 | 467 | | | | | | | | | | | | | | | **1,727** | 19,595 |
| 1969/70 | | | | 204 | 983 | 359 | | | | | | | | | | | | | | | **1,546** | 21,141 |
| 1970/71 | | | | 149 | 780 | | | | | | | | | | | | | | | | **929** | 22,070 |
| 1971/72 | | | | 93 | 740 | | 368 | | | | | | | | | | | | | | **1,201** | 23,271 |
| 1972/73 | | | | | 100 | | 518 | 178 | | | | | | | | | | | | | **796** | 24,067 |
| 1973/74 | | | | | | | 692 | 238 | 173 | | | | | | | | | | | | **1,103** | 25,170 |
| 1974/75 | | | | | | | 795 | 444 | 237 | | | | | | | | | | | | **1,476** | 26,646 |
| 1975/76 | | | | | | | 780 | 160 | 389 | 88 | | | | | | | | | | | **1,417** | 28,063 |
| 1976/77 | | | | | | | 727 | 100 | 473 | 344 | | | | | | | | | | | **1,644** | 29,707 |
| 1977/78 | | | | | | | 369 | | 356 | 245 | | | 5 | | | | | | | | **975** | 30,682 |
| 1978/79 | | | | | | | | | | | 492 | | 369 | | 150 | | 41 | | | | **1,052** | 31,734 |
| 1979/80 | | | | | | | | | | | 819 | | 358 | | 221 | | 168 | | 50 | | **1,616** | 33,350 |
| 1980/81 | | | | | | | | | | | 177 | 128 | 98 | 91 | 121 | 80 | 87 | 43 | 20 | 15 | **860** | 34,210 |
| 1981/82 | | | | | | | | | | | | 152 | | 254 | | 205 | | 102 | | 30 | **743** | 34,953 |
| 1982/83 | | | | | | | | | | | | | | 86 | | 36 | | 13 | | 12 | **147** | 35,100 |
| **TOTAL** | **8,803** | **150** | **5,414** | **2,364** | **5,446** | **826** | **4,249** | **1,120** | **1,628** | **677** | **1,488** | **280** | **830** | **431** | **492** | **321** | **296** | **158** | **70** | **57** | **35,100** | |

***During the expansion and construction period at Geelong Works, many dealerships throughout the country were also upgraded, built or relocated. The following are a few examples from 1956.***

**Bundaberg, Queensland**
Farmers Centre was located in a series of unattractive shops, at right, in 1946
Above —Business is conducted in the new premises in 1956

**Berri, South Australia**
Growers Services Ltd. were originally located in a small shop in the busy section of town. A vastly improved base of operations exists in 1956 with showroom, service parts and a service station.

**Windsor, New South Wales**
Condon's Truck and Tractor Service—
The old retail corner store has been replaced by a spacious premises by 1956.

**Narrabri, New South Wales**
Griffco IH dealers had a basic start in simple corrugated iron sheds that were built as a temporary base of operations during the construction of their modern premises.

**Mildura, Victoria**

Tractor and Implements Pty. Ltd. shows a delivery of 14 Farmall A tractors to farmers from their old premises. In 1956 they were located in the more modern building.

**Geelong, Victoria**

The original premises were located in a small garage. A more modern showroom, offices, parts department and service station were occupied by the dealer in 1956.

**Perth, Western Australia**
The expanse of glass greets visitors to C.C. & F. Leuba at Redcliffe, a suburb of Perth.

**Busselton, Western Australia**
Cowley & Davis cater to the users of IH products in the rich dairying and timber area

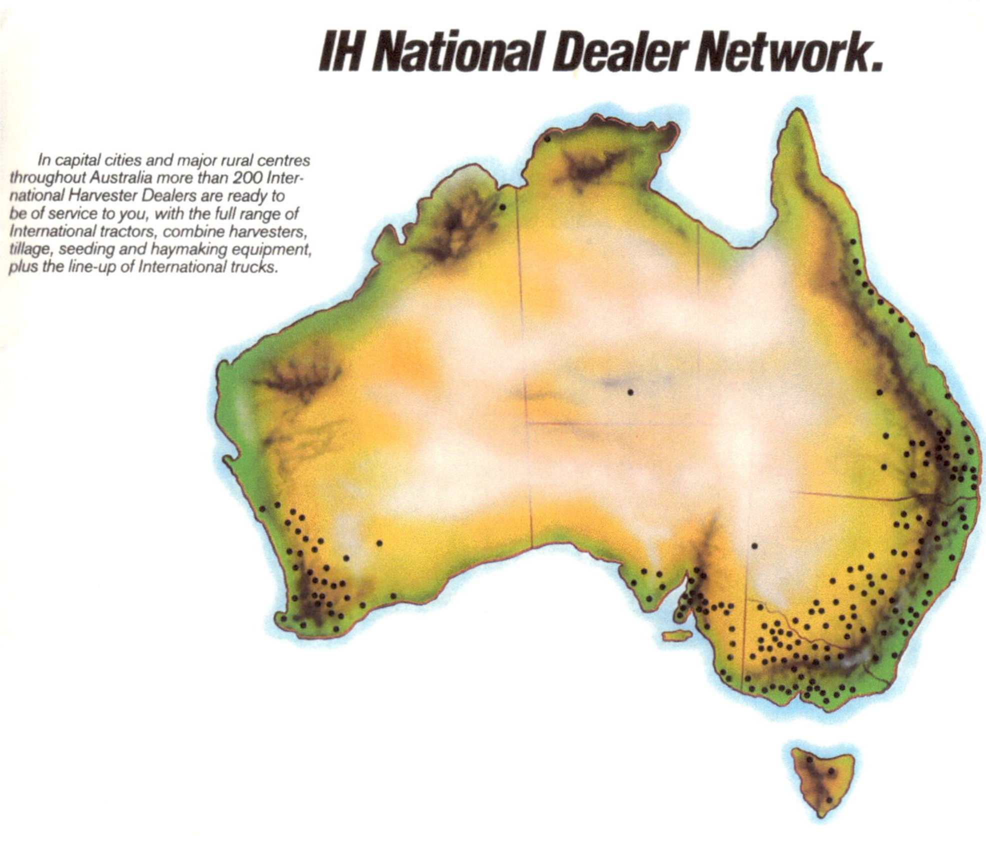


**1985 International Harvester Australia Network—Black dots represent each dealership.**

**Facts about Australia:**
Australia is the sixth largest country in the world. It's about the same size as the 48 mainland states of the USA and 50 per cent larger than Europe, but has the lowest population density in the world - only two people per square kilometer.

Australia's coastline stretches almost 50,000 kilometers and is linked by over 10,000 beaches, more than any other country in the world. More than 85 per cent of Australians live within 50 kilometers of the coast.

www.ingramcontent.com/pod-product-compliance
Lightning Source LLC
LaVergne TN
LVHW071248210926
848486LV00008B/365
* 9 7 8 0 6 4 6 5 6 0 3 7 3 *